Sound Engineer's
Pocket Book

D0223693

Sound Engineer's Pocket Book
Second Edition

Edited by
Michael Talbot-Smith

Focal Press
OXFORD AUCKLAND BOSTON JOHANNESBURG MELBOURNE NEW DELHI

Focal Press
An imprint of Butterworth-Heinemann
Linacre House, Jordan Hill, Oxford OX2 8DP
225 Wildwood Avenue, Woburn, MA 01801–2041
A division of Reed Educational and Professional Publishing Ltd

A member of the Reed Elsevier plc group

First published 1995
Reprinted 1998
Second edition 2000

© Reed Educational and Professional Publishing Ltd 2000

British Library Cataloguing in Publication Data
Sound engineer's pocket book. – 2nd ed.
   1. Acoustical engineering – Handbooks, manuals, etc.   2. Sound
  – Recording and reproducing – Handbooks, manuals, etc.
  I. Talbot-Smith, Michael
  621.3'828

Library of Congress Cataloguing in Publication Data
A catalogue record for this book is available from the Library of Congress

ISBN 0 240 51612 5

Typeset by Florence Production Ltd, Stoodleigh, Devon
Printed and bound in Great Britain by Biddles Ltd, *www.biddles.co.uk*

Contents

Preface to second edition   vii
Introduction   viii

1	Basic principles	1
2	The physics of sound waves	12
3	The hearing process	19
4	Acoustic noise and its measurement	23
5	Typical noise levels	28
6	Electromechanical analogies	31
7	Digital principles	37
8	Acoustics	41
9	Sound isolation	45
10	Microphones	53
11	Radio microphone frequencies	72
12	Loudspeakers	76
13	Stereo	84
14	Analogue sound mixing equipment	94
15	Signal processing	104
16	Analogue recording and reproduction	112
17	Analogue noise reduction	122
18	Compact disc	131
19	Digital audio tape	141
20	Audio measurements	154

21	Digital equipment	159
22	Midi	166
23	Studio air-conditioning	173
24	Distribution of audio signals	177
25	Radio propagation	182
26	Digital interfacing and synchronization	186
27	Ultrasonics	191
28	Radio studio facilities	195
29	Connectors	202
30	Public address data	205
31	Useful literature	207
	Index	211

Preface to second edition

I have had many gratifying comments about this Pocket Book. It's good to know that it fills a need, at least for some professionals. The second edition has been very carefully looked at and many alterations and (hopefully) improvements have resulted.

I particularly thank Keith Spencer-Allen for his critical study of the first edition and also for providing most of the material for the section on Digital Audio Tape. I'm also grateful to Vivian Weeks for his major contribution to the section on Audio Measurements.

Michael Talbot-Smith

Introduction

Some pocket books seem to be miniature text books. My view is that the place for text books is back in the office for occasional reference! Pocket books, I think, should contain the kind of information that may be needed frequently; that such a book has an important place in the pocket, and is as essential there as one's wallet or car keys.

Of course, it is unlikely that everyone is going to need all the information in such a volume, but my hope is that this Pocket Book will be important to most sound engineers most of the time.

For fuller information on any topic in it the reader is referred to the *Audio Engineer's Reference Book* (Focal Press, second edition, 1999) from which most, but not all of the data has been drawn. Some of the material comes from my own work and some comes from the excellent little *Audio System Designer* handbook, published a few years ago by Messrs Klark–Technik. I gratefully acknowledge their permission to quote from it.

The ideal target, as I have said, would have been attained if any practising engineer underwent a panic attack if he or she found that they had left the relevant pocket book behind. I doubt if that is ever going to be quite the case, but perhaps this pocket book for sound engineers might approach that ideal.

Michael Talbot-Smith

1 Basic principles

Useful relationships

Velocity of light in free space $= 299\ 792.458$ km/s

Surface area of a sphere $= 4\pi r^2$

Volume of a sphere $= \dfrac{4}{3}\pi r^3$

Peak value of sine wave $= 1.414 \times$ r.m.s.

$e =$ base to natural logarithms $= 2.718\ 28$

SI Prefixes and multiplication factors

Multiplication factor			Prefix	Symbol
1 000 000 000 000	=	10^{12}	tera	T
1 000 000 000	=	10^{9}	giga	G
1 000 000	=	10^{6}	mega	M
1 000	=	10^{-3}	kilo	k
1/1000	=	10^{-3}	milli	m
1/1 000 000	=	10^{-6}	micro	μ
1/1 000 000 000	=	10^{-9}	nano	n
1/1 000 000 000 000	=	10^{-12}	pico	p
		10^{-15}	femto	f
		10^{-18}	atto	a

Definitions
Electric current (*I*). The unit is the ampere (A). One ampere is the current which when flowing in two infinitely long conductors, one metre apart and of negligible cross-section, produces a force between them of 2×10^{-7} newtons per metre length.

Force (*F*). Unit the newton (N). A force of one newton will cause a mass of one kg to accelerate by one m/s/s.

Work or energy (W). Unit is the joule (J). One joule of work is done when one newton of force moves through one metre in the direction of the force.

Power (P). The unit is the watt (W). Power is the rate at which work is done: one joule of energy expended in one second requires one watt of power. (The non-preferred unit of the *horse power* is equal to 746 watts.)

Charge (Q). Unit is the coulomb (C). It is the quantity of electricity passing through a conductor when one ampere flows for one second.

Potential difference (V). Unit is the volt (V). One volt of potential difference exists between two points if one joule of work is done in transferring one coulomb of charge between them.

Resistance (R). Unit the ohm (Ω). One ohm is that resistance in which a current of one ampere generates energy at the rate of one joule per second.

Capacitance (C). Unit is the farad (F). A capacitor with a capacitance of one farad will have the potential between its plates increased by one volt when the charge on them is increased by one coulomb.

Magnetic flux (Φ). Unit is the weber (Wb). If the flux experienced by a conductor changes at the rate of one weber per second then a potential difference of one volt is produced across the ends of the conductor.

Flux density (B). The unit is the tesla (T) (Wb/m²). Flux density is the flux per square metre at right angles to the field.

Inductance (L). Unit is the henry (H). A closed circuit has an inductance of one henry when a potential of one volt is produced by a rate of change of current of one ampere per second.

Pressure (p). The unit is the pascal (Pa). This is equivalent to one newton per square metre (N/m²).

Frequency (*f*). The unit is the hertz (Hz). The number of complete waves passing a fixed point in one second.

(The unit cycles/second, c/s or c.p.s, is not currently approved.)

Wavelength (λ). The distance between corresponding points on successive waves.

Permeability (μ).

1. The *permeability of free space* (μ_0) is the ratio of the flux density (*B*) to the magnetizing force (*H*) in a non-magnetic material (space).

   $$\mu_0 = B/H = 4\pi \times 10^{-7} \text{ (H/m)}$$

2. The *relative permeability* (μ_r) of a material is the factor by which the flux density increases for the same magnetizing force.

Bases in mathematics

Base 10 ('decimal' or 'denary')

$$\text{e.g. } 396_{10} = (3 \times 10^2) + (9 \times 10^1) + (6 \times 10^0)$$
$$= 300_{10} + 90_{10} + 6_{10}$$
$$= 396_{10}$$

Base 2 (binary)

$$\text{e.g. } 1011_2 = (1 \times 2^3) + (0 \times 2^2) + (1 \times 2^1) = (1 \times 2^0)$$
$$= 8_{10} + 0 + 2_{10} + 1_{10}$$
$$= 11$$

Base 16 (hexadecimal)

i.e. 0,1,2,3,4,5,6,7,8,9,A,B,C,D,E,F

Hexadecimal numbers are normally prefixed by '&'.

e.g. $\&B2F_{16}$ $= (11 \times 16^2) + (2 \times 16^1) + (15 \times 16^0)$

$= 2816_{10} + 32_{10} + 15_{10}$

$= 2863_{10}$

The decibel

Basically the decibel is a unit of comparison of two powers. If the powers are P_1 and P_2 then the gain or loss of one with respect to the other is

$$dB = 10 \log_{10}(P_1/P_2)$$

(Note that the subscript indicating that the logarithm is to the base 10 is normally omitted.)

Since Power $\propto$ (voltage)^2 and also to (current)^2 when used with voltages or currents (or pressures in sound) the expression becomes

$$dB = 20 \log (V_1/V_2)$$

or $$dB = 20 \log (I_1/I_2)$$

or $$dB = 20 \log (p_1/p_2)$$

where p_1 and p_2 are sound pressures

Acoustic resonances

Membrane or panel.

$$f_0 = \frac{60}{\sqrt{Md}}$$

where M is the mass/unit area of the membrane or panel in kg/m^2 and d is the depth of the air space in metres.

Helmholtz resonator

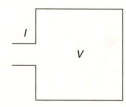

(Helmholtz resonator, continued)

$$f_0 = \frac{cr}{2\pi} \sqrt{\frac{2\pi}{V(2I + \pi r)}}$$

where c is the velocity of sound, r is the radius of the neck, l is the length of the neck and V is the volume.

Decibel values for ratios of power, current and sound pressure level (SPL)

Power ratio	Voltage, current or SPL ratio	dB
1	1	0
1.5	1.23	1.8
2.0	1.41	3.0
2.5	1.58	4.0
3.0	1.73	4.8
4.0	2.00	6.0
5.0	2.24	7.0
10.0	3.16	10.0
20.0	4.47	13.0
30.0	5.48	14.8
50.0	7.07	17.0
100.0	10.00	20.0
500.0	22.36	27.0
1 000.0	31.6	30.0
10 000.0	100.0	40.0

Electrical formulae

Ohm's Law:

$$V = IR$$

$$I = \frac{V}{R}$$

$$R = \frac{V}{I}$$

Power:

$$P = IV = I^2R = \frac{V^2}{R} \text{ Watts}$$

Resonance:

$$f_0 = \frac{1}{2\pi\sqrt{LC}}$$

in series LC circuits and in parallel circuits where resistance is small and can be neglected.

Resistor colour code

Bands 1 and 2: first and second digits.

Band 3: multiplier. The colour code is the same as for bands 1 and 2 but indicates the number of zeros after the first two numbers.

Band 4: percentage tolerance (if present).

Band 5: stability. Pink, if present, indicates high stability.

Bands 1 and 2:		
	Black	0
	Brown	1
	Red	2
	Orange	3
	Yellow	4
	Green	5
	Blue	6
	Violet	7
	Grey	8
	White	9

Band 4 (tolerance):	No band	± 20%

Silver	± 10%
Gold	± 5%
Red	± 2%

Amplitude modulation

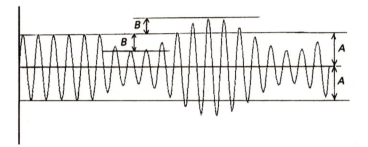

Figure 1 Modulation index is given by $m = B/A$

Frequency modulation

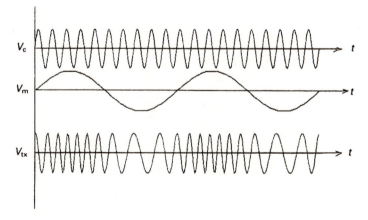

Figure 2 Frequency modulation

(Continued on page 11)

Decibel values for power ratios

Power ratio	dB
1	0
1.5	1.76
2.0	3.01
2.5	3.98
3.0	4.77
4.0	6.02
5.0	6.99
6.0	7.78
7.0	8.45
8.0	9.03
9.0	9.54
10.0	10.00
20.0	13.01
30.0	14.77
40.0	16.02
50.0	16.99
60.0	17.78
70.0	18.45
80.0	19.03
90.0	19.54
100.0	20.00
200.0	23.01
300.0	24.77
400.0	26.02
500.0	26.99
600.0	27.78
700.0	28.45
800.0	29.03
900.0	29.54
1 000.0	30.00
5 000.0	36.99
10 000.0	40.00
100 000.0	50.00
1 000 000.0	60.00

Decibel values for voltage, current and sound pressure level (SPL)

Voltage, current or SPL ratio	dB
1	0
1.5	3.52
2.0	6.02
2.5	7.96
3.0	9.54
4.0	12.04
5.0	13.98
6.0	15.56
7.0	16.90
8.0	18.06
9.0	19.08
10.0	20.0
20.0	26.02
30.0	29.54
40.0	32.04
50.0	33.98
60.0	35.56
70.0	36.90
80.0	38.06
90.0	39.08
100.0	40.0
200.0	46.02
300.0	49.54
400.0	52.04
500.0	53.98
600.0	55.56
700.0	56.90
800.0	58.06
900.0	59.08
1 000.0	60.00
5 000.0	73.97
10 000.0	80.00
100 000.0	100.00
1 000 000.0	120.00

Voltage, current or SPL ratios for decibel values

dBs	Voltage, current or SPL ratios
0	1.00
1.5	1.19
2.0	1.26
2.5	1.33
3.0	1.41
3.5	1.50
4.0	1.58
5.0	1.78
6.0	2.00
7.0	2.24
8.0	2.51
9.0	2.82
10.0	3.16
15.0	5.62
20.0	10.00
25.0	17.78
30.0	31.62
35.0	56.23
40.0	100.00
45.0	177.82
50.0	316.23
55.0	562.34
60.0	1 000.0
65.0	1 778
70.0	3 162
75.0	5 623
80.0	10 000
85.0	17 782
90.0	31 622
95.0	56 234

(Frequency modulation, continued from page 7)

The modulation index for an f.m. system is given by

$$m = \frac{\text{carrier frequency deviation } f_D}{\text{modulating frequency } f_m}$$

2   The physics of sound waves

Units
Pressure:

Pascal (Pa)   =   1 N/m² (also equal to 10 dyne/cm²)

Bar: 1 bar   = 10^5 Pa

Torr: 1 torr   =   133.22 Pa

Sound wave pressures fall in the range between 0.00002 Pa or 20 µPa), corresponding approximately to the average ear's threshold at around 3–4 kHz, to about 200 Pa, generally reckoned to be about the pain level.

Intensity
Watts/m² (W/m²), or, more practical in sound, µW/m².

Velocity of sound, typical values
See table opposite.

A general expression for the velocity of sound in gases is given by:

$$c = (\gamma P/\rho)^{-2}$$

where γ is the ratio of the specific heats of the gas (1.414 for air), P is the pressure and ρ is the density of the gas.

Frequency (f) and wavelength (λ)
These are related by the expression

$$c = f\lambda$$

Note that this formula applies to all waves. In the case of electromagnetic waves (radio, light etc.) c is approximately 300 000 km/s (3×10^8 m/s) as opposed to about 340 m/s for sound waves in air. The formula could be used only with caution for surface waves on water as the velocity of the waves can vary with amplitude.

Velocity of sound, typical values

Substance	c (m/s)
Air, 0°C*	331.3
Hydrogen, 0°C	1284
Oxygen, 0°C	316
Carbon monoxide, 0°C	337
Carbon dioxide, 18°C	266
Water, 25°C	1498
Sea water, 20°C	1540
Glass	~5000
Aluminium	5100
Brass	3500
Copper	3800
Iron (wrought)	5000
Iron (cast)	4300
Concrete	3400
Steel	5000–6000
Wood, deal, along grain	5000
oak	4000–4400
pine	3300

* The velocity of sound in air increases with temperature by approximately 2/3 of a m/s per °C rise in temperature. More accurately: $c = 331 + 0.6t$ where t is the temperature in °C.

Sound wavelengths in air at 20°C

Frequency (Hz)	Wavelength (m)
16	21.43
20	17.15
30	11.43
50	6.86
100	3.43
200	1.72
500	0.69
1 000	0.34
5 000	0.069
10 000	0.034
16 000	0.021

The inverse square law

Intensity (I) falls off with distance (d) according to:

$$I \propto \frac{1}{d^2}$$

Pressure, on the other hand, follows the law:

$$P \propto \frac{1}{d}$$

The Doppler effect

Assuming a stationary medium, if the source is moving towards the observer with velocity v_s then the apparent frequency, f_a, is given by

$$f_a = \frac{f_c}{(c - v_s)}$$

If the observer is moving towards the source with velocity v_o then

$$f_a = \frac{f(c + v_o)}{c}$$

The musical scale

The table opposite shows the *equal-tempered scale*, where the ratio of the frequency of one note to the next one above it is $\sqrt[12]{2}$ or 1.059 463 1.

Frequencies of vibrations in pipes and strings

Pipe open at one end:

$$f = \frac{nc}{4(l + a)}$$

where $n = 1,2,3$ etc. l is the length of the pipe and a is the end correction.

For a pipe with a significant flange a is roughly $0.8r$, where r is the radius.

(Continued on page 17)

The musical scale

Note	Frequency (Hz)
A	220.00
A#	233.08
B	246.94
C	261.63
C#	277.18
D	293.66
E	311.13
E#	329.63
F	349.23
F#	369.99
G	392.00
G#	415.30
A	440.00
A#	466.16
B	493.88
C'	523.25

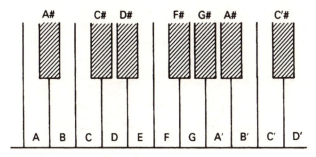

Figure 3 Portion of a keyboard

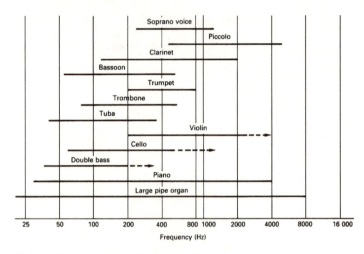

Figure 4 Frequency ranges of some typical sounds

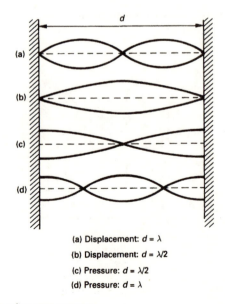

(a) Displacement: $d = \lambda$
(b) Displacement: $d = \lambda/2$
(c) Pressure: $d = \lambda/2$
(d) Pressure: $d = \lambda$

Figure 5 Standing wave patterns

(Frequencies of vibrations in pipes and strings, continued from page 14)

In the case of an unflanged pipe a is about $0.6r$.

Pipe open at both ends:

$$f = \frac{nc}{2(l + 2a)}$$

The velocity of a transverse wave in a string of tension T and mass m per unit length:

$$c = \left(\frac{T}{m}\right)^{1/2}$$

The lowest frequency of vibration in such a string is then

$$f = \frac{1}{2l}\left(\frac{T}{m}\right)^{1/2}$$

Some Italian terms found in music

fff (Molto fortissimo)	extremely loud
ff (fortissimo)	very loud
f (forte)	loud
mf (mezzo forte)	fairly loud
mp (mezzo piano)	fairly quiet
p (piano)	quiet
pp (pianissimo)	very quiet
ppp (molto pianissimo)	extremely quiet
meno	less
più	more
staccato	a very short sound
crescendo	becoming louder
diminuendo	becoming quieter
Grave	very slowly
Lento	slowly
Largo	broadly
Larghetto	rather broadly
Adagio	in a leisurely manner
Andante	at a moderate walking pace
Moderato	at a moderate speed

Allegretto	fairly fast
Allegro	fast
Vivace	lively
Presto	very fast
Prestissimo	as fast as possible
*pizzicato**	the strings are plucked
*arco**	played with the bow
	(*stringed instruments only)
sempre	'always' or 'continue' – as a reminder that an earlier instruction to play in a particular way is still in force.

3 The hearing process

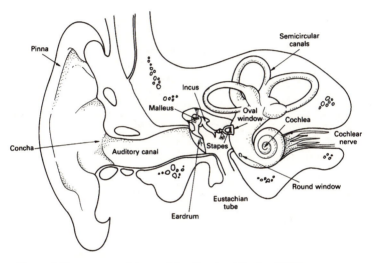

Figure 6 The structure of the ear (after Kessel and Kardon (1979))

The response of the ear to different frequencies
This is best shown graphically, as in Figure 8. Each line is an *equal loudness curve*. See below for an explanation of Phons.

The reference zero, a sound pressure level (SPL) of 0 dB, is taken to be a pressure of 20 µPa (0.000 02 N/m²).

Frequency discrimination
The smallest change in frequency that can be detected is known as the *frequency difference limen, dl*. At sound levels 20 dB above the threshold, and for frequencies below 1–2 kHz, and for durations

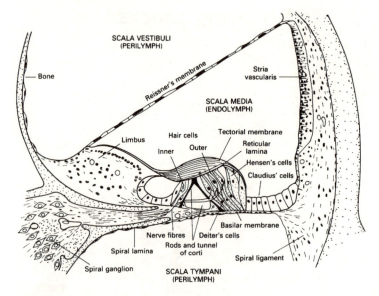

Figure 7 Cross-section through one turn of the cochlea

greater than 0.25s dl is reasonably constant at 1 Hz. Above 2 kHz dl increases approximately in proportion to the frequency.

Loudness

Loudness is a subjective quality related to sound intensity. It is generally accepted that with steady tones a change of level of 1 dB can just about be detected. With speech and music 3dB is the minimum change of level that is normally detectable.

In general a change of level of 10 dB causes a doubling or halving of the perceived loudness.

The sone. This is an attempt to have a unit which is proportional to loudness. It is defined on the basis of one sone being the loudness at 40 dB above threshold, with 1 sone added for each 10 dB increase in level (or halved for each 10 dB below 40 dB).

$$s = 2^{(p-40)/10}$$

where s is the number of sones for a sound pressure of p dB above the threshold.

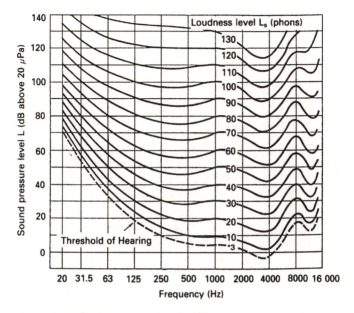

Figure 8 Equal loudness curves (from Robinson and Dadson (1956))

The phon. Phons are defined by the SPL in dB at 1 kHz and the Equal Loudness Curves as shown in Figure 7. Thus, taking for example the 80 phon curve, at 63 Hz the SPL has to be about 90 dB to sound as loud as 80 dB at 1 kHz.

Noise masking
The property of a louder signal to render inaudible a quieter sound. This effect is greatest when the two sounds are within the same critical band and when the masking signal is lower than the masked one. Figure 9 shows the effect of a 90 Hz-wide band of noise centred on 410 Hz on sounds of other frequencies.

When the masking signal's level is 60 dB it will provide about 30 dB of masking for frequencies between about 350 Hz and 600 Hz. With a masking noise level of 80 dB there will be roughly 20 dB of masking between 260 Hz and 1500 Hz. (Courtesy of Egan and Hale (1950).)

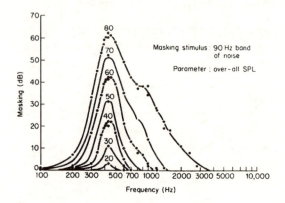

Figure 9 Noise masking

Beats

With two tones whose frequencies are close together the signals
will go alternately in and out of phase. One of three sensations are
then possible.

> up to about 10 Hz difference the low frequency alternations
> in level are perceived;

> between frequency differences of about 10 Hz and about
> 50–500 Hz (according to the mean frequency) there is a sensa-
> tion of 'roughness';

> at higher frequencies the two frequencies are heard sepa-
> rately.

These beats are not difference tones and do not imply a non-
linearity in the hearing process.

Difference tones, can however, be heard when the two frequencies are in the ratio 1.1:1 or 1.2:1 up to 1.5:1.

4 Acoustic noise and its measurement

Definitions
Noise level meters, unless very cheap, have primarily two settings, giving readings in dBA and LIN (or dBC).

dBA or dB(A) – either seems currently acceptable in the UK. This is a measurement which gives a reasonable approximation to loudness as perceived by the average person. The 'A' weighting network gives the measuring instrument a response which allows for the ear's lack of sensitivity to low frequencies. See Figure 10.

dB(C). Some noise level meters, particularly in the USA, have a setting marked 'C'. (Older meters may have a 'B' setting. This represents a weighting network which is between 'A' and 'C' characteristics. This has fallen into disuse.)

LIN. Flat response.

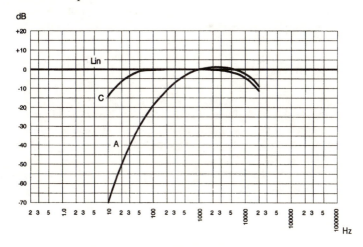

Figure 10 A, C and LIN frequency weightings

$L_{eq,T}$ – equivalent noise level. This is the sound level which, integrated over a period of time, T, is equivalent to a steady SPL.

$$L_{eq,T} = \log_{10}\left\{\left[\frac{1}{T}\int_{t_1}^{t_2}p^2(t)\,\mathrm{d}t\right]\bigg/p_0^2\right\}\mathrm{dB}$$

If, as is usual, measurements are made with the A weighting network then this is indicated by the subscript A: $L_{Aeq,T}$.

$L_{Aeq,T}$ is used widely in assessing the likely effect of industrial, and other noise sources such as roads, on residential areas.

The British Standard BS 4142: 1990 ('Rating industrial noise affecting mixed residential and industrial areas') should be consulted for more information. See the note on page 27.)

Other terms found in various aspects of noise measurement are:

L_n where n is a number, often 10, 50 or 90, meaning that the SPL indicated by L is exceeded for n per cent of the time. Thus if L_{90} were 60 dBA the value of 60 dBA would be exceeded for 90 per cent of the time under consideration.

$L_{eP,d}$ is very similar to $L_{Aeq,T}$ in that essentially the same type of equipment is used to measure it. It represents, however, the personal daily 'dose' of noise. In effect, then, $L_{Aeq,T}$ applies to locations, $L_{eP,d}$ applies to people.

Hearing damage

The sound levels which are a risk to a person's hearing can be a subject of some controversy, especially as it seems that different individuals may have different susceptibilities. However it is generally accepted in the UK and much of Europe that there is little risk of hearing damage if the ears are exposed to an L_{Aeq} of 90 dB for 8 hours each day.

Above 90 dB the permissible exposure time is halved for each 3 dB increase. See table opposite.

Permissible exposure times

L_{Aeq}	Permissible exposure time
90	8 hours
93	4 hours
96	2 hours
99	1 hour
102	30 mins
105	15 mins
108	8 mins
111	4 mins

Current health and safety regulations

In the UK these can be summarized by stating that there are two Action Levels:

1. If the $L_{eP,d}$ for the employee(s) exceeds 85 dBA but is less than 90 dBA then the employer must inform the employee(s) and ensure that suitable hearing protection is available to all who ask for them.

2. Above 90 dBA or when the peak sound pressure exceeds 200 Pa (140 dBA) the employer must reduce noise levels as far as is practicable by means other than ear protectors and mark all zones of noise levels higher than 90 dBA.

The above is the barest of outlines and HSE literature should be consulted for more information.

Sound level meters

There are four defined grades:

Grade 0: the very highest standard and mainly used for laboratory work;

Grade 1: for precision measurements;

Grade 2: for general purposes;

Grade 3: for sound surveys.

Most meters except perhaps Grade 3 will have the ability to measure the following:

'Impulse' or 'Peak' readings

'Fast', which averages over ⅛ s

'Slow', which averages over 1 s.

Noise measurements in the field should normally be made with at least 1.2 m of clear space (including the ground) all round the microphone. The noise being measured should be at least 10 dB above background noise. If not a correction should be applied as in the table below.

Noise level reading, $L_{Aeq,T}$ minus background, $L_{A90,T}$ dB	Correction: subtract from noise level reading dB
6 to 9	1
4 to 5	2
3	3
<3	Difficult to correct

Addition of noise levels

Noise levels in dB cannot meaningfully be added together, as, being logarithms the result will be a multiplication.

Antilogs of the readings must be used, as shown below. Note that the noise reading should be divided by 10 before finding the antilog as antilogs of 100 or more will result in an ERROR indication on normal calculators. To take an example:

Find the sound level resulting from 92 dBA + 98 dBA + 96 dBA + 68 dBA

Noise reading	Divide by 10	Antilog
92	9.2	1.585×10^9
98	9.8	6.310×10^9
96	9.6	3.981×10^9
68	6.8	6.310×10^6
	SUM	1.188×10^{10}

Log $(1.188 \times 10^{10}) = 10.075$

Result $= 10 \times 10.075 = 100.75$ dBA

(Note that the contribution of the reading of 68 dBA is insignificant, being some 30 dB below the other readings.)

The above method, adjusted to take time durations into account, may be used to find an $L_{eP,d}$ from a series of measurements of L_{Aeq}.

National and international standards

In the list below British Standards relating to acoustic noise and its measurements are given with equivalent standards in **bold**.

The following abbreviations are used:

ANSI   American National Standards Institute

BS   British Standards Institute

IEC   International Electrotechnical Commission

ISO   International Standardization Organization

BS 3539: 1986
Sound level meters for the measurement of noise emitted by motor vehicles
*BS 4142: 1990
Method for rating industrial noise affecting mixed residential and industrial areas
BS 5969: 1981 **(IEC 651: 1979)**
Specification for sound level meters
BS 6402: 1994 **(IEC 1252: 1993)**
Personal sound exposure meters
BS 6698: 1986 **(IEC 804: 1985)**
Integrating–averaging sound meters
BS 7580: 1992 **(IEC 645–1: 1992)**
Specification for the verification of sound level meters
ANSI S1.4 – 1983
Specification for sound level meters
ANSI S1.25 – 1991
Specification for personal dosemeters

* Most of BS 4142: 1990 is relevant to any outdoor measurement. It can be regarded as a safe procedure to use in many applications additional to industrial noise, such as traffic, entertainment and animal sounds, such as kennels.

5 Typical noise levels

The measurements given below must not be taken as definitive since there can be significant variations between different circumstances: for instance traffic noise can vary with the type and speed of vehicles, wear of bearings in industrial tools can alter the noise emission. Nevertheless the figures given are indicative of the approximate sound levels which might be likely in the stated conditions. Unless otherwise stated the measurements were made at 1 m from the noise source.

Source	dBA	Notes
Transport		
Jet aircraft	140	At 30 m, take-off
Jet aircraft, small	120	At 150 m
Train, freight	99	Wagons at 12 m
Train, freight	96	Locomotive at 12 m
Train, high speed	95	At 12 m from track
Train interior	75	Medium size train at 70 m.p.h
Road, medium busy	80	At roadside
Road, medium busy	52	At 150 m
Road, medium busy	49	At 300 m
Motorway	48	3 km distance. Little wind.
Industrial machinery		
Air line	104	1 m from nozzle
Angle grinder	105	On 4 mm steel
Automatic saw	99	1 m from saw blade
Belt sander, joinery	87	On soft wood
Brazing	94	At operator's ears

Source	dBA	Notes
Gas welding torch	100	
Guillotine	85–90	At operator's ears
Hammer, 2 lb	120	On 6 mm mild steel, on anvil
Hammer, 1½ lb	116	On tinplate on anvil
Hammer, 2 lb	116	On steel pipe on anvil
Handgrinder	98	On box steel
Handgrinder	110	On 9 mm sheet steel
Induction furnace	85	At 3 m
Industrial vacuum cleaner	94	
Large fan	112	Used in glass annealing
Lathe	85	
Milling machine	80	3000 r.p.m
Polishing wheel	72	On brass
Power saws	100–115	
Power presses	95–110	

Agricultural

Source	dBA	Notes
Chain saw	108	Fast, off load
Diesel Rotavator	97	At operator's ears
Digger	87	At driver's ear, PTO speed
Digger	93	1 m at side of engine
Dumper truck, 1 tonne	89	Driver's position, normal speed
Dumper truck, 1 tonne	100	1 m at rear, normal speed
Fork Lift Truck	85	Driver's position, working revs
Petrol Rotavator	88	At operator's ears
Tractors	80–90	Inside cab, medium speed
Tractors	70–80	Inside cab, engine tickover

Domestic

Source	dBA	Notes
Electric drill	100	½″ masonry drill into brick
Jigsaw, electric	97	Operator's ears, large area of wood, 30 mm thick
Loud radio or TV	70	
Normal speech	60–65	

Source	dBA	Notes
Petrol-engined rotary lawnmower	90	At operator's ears
Smoke alarm	105	
Miscellaneous		
Orchestra, fortissimo	100	At 5 m
Church bells	72	35 m from base of tower
Sea	68	20 m from breakers, Force 3 off-shore wind
Rural area at night	<30	No roads in the vicinity

6   Electromechanical analogies

Summary of analogue quantities, units, symbols and relationships

The table below gives the overall units and dimensions:

Mechanical	Acoustic	Electrical
Mass, M (kg)	Mass, M_a, Inertance	Inductance, H, (Henry)
Force, F (N)	Sound pressure, p	Volts (V)
Displacement, x (m)	–	Charge Q (C)
Velocity, $\dot{x}$ (m/s)	Volume velocity	Current $(Q/s, A), i$
Energy, work (J)	joules (J)	joules (J)
Power $(W, J/s)$	watts (W)	watts (W)
Stiffness, F/x	–	–
Compliance, x/F	Compliance, C_a	Capacitance, C
Frequency, f (I/s, Hz)	–	–
Resistance, R_m	Acoustic resistance, R_a	Resistance R
Temperature (K)	–	Magnetic flux density, B
Length, L	–	–

Impedance relationships

Mechanical impedance, $Z_m = F/\dot{x}$

Specific acoustic impedance, $Z_s = p/\dot{x}$

Analogous acoustic impedance $= p/\dot{x}A$, where A is the area of the element and p is sound pressure

Admittance $= 1/Z$

Transformations: $F = Bli$; $v = Bl\dot{x}$

Newton's laws: $F = M\ddot{x}$

Kinetic energy $= \frac{1}{2} M\ddot{x}^2$

Potential energy $= \frac{1}{2} sx^2$

The following table is more complete than the one on page 31 and includes magnetic units.

Analogue quantity	Symbol	Units in which it is expressed
Energy		joules $\equiv$ newton metre $\equiv$ coulomb volts
Power	W	watts = joule seconds = ampere volts
Charge	Q	coulomb metre
Mass	M	kilograms = newton seconds^2/metre
Potential	V	volts
Current	I	amperes = coulombs/second
Transduction coefficient	T	newtons/ampere = volts seconds/metre
Magnetic flux	ϕ	webers = volt seconds
Magnetic flux density	B	webers/metre
Magnetic field vector	H	amperes/metre
Magnetization	M	amperes/metre
Magnetomotive force	F	ampere turns = coulombs/s
Magnetic reluctance	R	ampere turns/weber = coulombs/volt seconds^2
Inductance	L	henries = webers/ampere = volt second^2/coulomb
Susceptibility	κ	
Electric impedance	Z	ohms
Electric resistance	R	= volts/ampere
Electric reactance	X	= volts seconds/coulomb
Force	F	newtons = joules/metre = coulomb volts/metre
Pressure	P	pascals = newtons/metre^2 = joules/metre^3
Density	ρ	kilograms/metre^3 = Ns^2/m^4
Stiffness	s	newtons/metre = coulomb volts/metre

Analogue quantity	Symbol	Units in which it is expressed
Compliance	C_m	metres/newton = metres^2/coulomb volt
Specific acoustic impedance	ρ_c	kilogram/metre^4 second = coulomb volts/metre^4 = newton seconds/metre^3
Mechanical impedance	z_m	newton seconds/metre = kilograms/second
Acoustic impedance	Z_A	newton seconds/metre^5 = kilograms/metre^4 second = ohms (acoustic)
Permittivity of free space	ϵ_0	8.854×10^{-12} = $(1/36\pi) \times 10^{-9}$ farads/m
Electric field	E	volts/metre
Electric displacement		coulombs/metre^2
Electric polarization		coulombs/metre^2
Capacitance	C	farads = coulombs/volt

Figure 11 shows the acoustic elements and their electrical equivalent circuits.

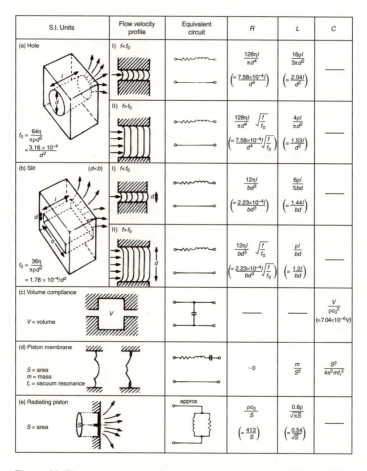

Figure 11 The acoustic elements, electrical equivalent circuits and R, L and C expressed in terms of the most readily measurable parameters. Viscosity coefficient $\eta = 1.86 \times 10^{-5}$ kg/(m/s), air density $\rho = 1.2$ kg/m³, velocity of sound $c_0 = 340$ m/s (Poldy 1988)

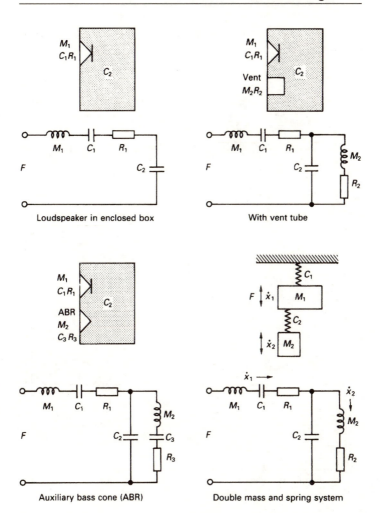

Loudspeaker in enclosed box

With vent tube

Auxiliary bass cone (ABR)

Double mass and spring system

Figure 12 Systems with two degrees of freedom

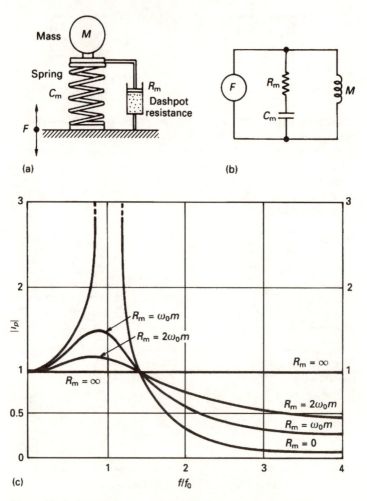

Figure 13 Analysis of vibration isolation. **(a)** damped mechanical system; **(b)** equivalent electrical circuit; **(c)** isolation factor (i_p) plotted against f/f_0 for various degrees of damping

7 Digital principles

Sampling rate

The number of times each second that the signal is 'measured'. A figure of 2.2 times the highest audio frequency is considered to be practical. The sampling rate for standard digital audio such as compact discs is 44.1, giving the highest audio frequency that can be recorded and replayed as 44.1/2.2 or just over 20 kHz. 48 kHz is an alternative 'professional' rate.

Bit rate is the number of 1s and 0s transmitted or recorded per second and is the product of the number of bits in the sample and the sampling frequency. Thus with a 16 bit system and a sampling rate of 44.1 kHz the number of bits per second is 705.6×10^3.

The number of bits/sample is closely related to the number of quantizing levels – the 'resolution' of the system as shown in the following table.

Quantizing levels	Number of bits
1 024	10
2 048	11
4 096	12
8 192	13
16 384	14
32 768	15
65 536	16
131 072	17
262 144	18
524 288	19
1 048 576	20
2 097 152	21
4 194 304	22
8 388 608	23
16 777 216	24

Each extra bit doubles the number of quantizing levels and thus halves the noise component. This means that there is a 6 dB improvement in signal-to-noise ratio for each bit added.

A reasonably good approximation to the signal-to-noise ratio in terms of the number of bits is given by

$$\frac{S}{N} = 6N + 1.75 \text{ dB}$$

The following table of S/N ratio for different numbers of bits uses the above formula, rounded to the nearest dB.

Number of bits	Signal/noise ratio dB
10	62
11	68
12	74
13	80
14	86
15	92
16	98
17	104
18	110
19	116
20	122
21	128
22	134
23	140
24	146

Definitions

1. **Dither.** The addition of a low level random noise signal to reduce quantizing noise.

2. **Two's complement.** A way of dealing with positive and negative numbers without resorting to + and − signs. In a digital system the most significant bit (MSB) represents the sign, being 1 for a negative number and 0 for a positive number. Using 10-bit numbers as examples the positive

range is from 0000000000 to 0111111111 while the negative range is from 1111111111 to 1000000000. As a signal crosses the zero line all the bits change from 1s to 0s and vice versa.

3. **Aliasing.** The production of spurious frequencies arising from beats formed by the sampling frequency and out-of-band frequencies. These can be eliminated by filtering although very steep 'brick wall' filters may be needed (See 'Oversampling' below.)

4. **Oversampling.** The data is read two, four, eight etc. times the sampling frequency. This has the effect of raising the aliasing frequencies by one, two, three etc. octaves, thus simplifying the design of filters.

Digital signal processing

Mixing – the digital equivalent is the addition of the numbers representing the samples.

Gain. In the digital domain this means multiplication of the numbers. Thus a 6 dB gain is the equivalent of multiplying by 2. Division by 2 would mean a gain reduction of 6 dB.

Error detection and correction

The simplest method of error detection is called 'parity'. The number of 1s in the sample is made up when necessary to an even number by the addition of an extra 1. (This is 'even' parity: 'odd' parity is sometimes used and then the number of 1s is made up to an odd number.)

Parity fails if there are more than two errors, but in a well-engineered system this should be a very rare occurrence.

On detection of an error the following are possible options:

1. correct the error;

2. conceal the error by, for example, repeating the previous sample;

3. muting the error.

More complex parity methods make it possible to tell which bit(s) are wrong and thus correct them.

To illustrate such a process:

Original word (12-bit for simplicity):

b_1	b_2	b_3	b_4	b_5	b_6	b_7	b_8	b_9	b_{10}	b_{11}	b_{12}

Rearranged:

b_1	b_2	b_3	b_4	P_1
b_5	b_6	b_7	b_8	P_2
b_9	b_{10}	b_{11}	b_{12}	P_3
P_4	P_5	P_6	P_7	

The parity bits P_1 etc. can check on each column and each row. Thus if errors are shown by P_2 and P_4 it is clear that b_5 is faulty – if it appears as a 1 it is changed to a 0, and so on.

8 Acoustics

Resonant frequencies in rooms

Rayleigh's formula for the frequencies of modes in a rectangular room:

$$f_m = \frac{c}{2\sqrt{(n_L/L)^2 + (n_W/W)^2 + (n_H/H)^2}}$$

where n_L, n_W and n_H are positive integers, including 0, L, W and H are the room dimensions and c is the velocity of sound.

Modal density. If the number of resonant modes N below the frequency f is fairly high then a reasonable approximation is given by

$$N \approx 4\pi f^3 \frac{V}{3c^3} + \pi f^2 \frac{S}{4^2} + \frac{fL}{8c}$$

where V is the volume,
S is the total surface area, $2(LW + WH + HL)$ and
L is the sum of the edge lengths $4(L + W + H)$.

Reverberation time (RT, r.t. or T_{60})

This is defined as the time taken for the reverberant sound level to decay through 60 dB. An approximate expression, the so-called Sabine formula, valid for rooms with small amounts of absorption is:

$$T_{60} = 0.16 \frac{V}{S\bar{a}}$$

where V is the volume in m^3, S is the total surface area in m^2 and $\bar{a}$ is the average absorption coefficient of the surfaces.

A more accurate formula, applicable in rooms with large amounts of absorption is due to Eyring and Norris:

$$T_{60} = \frac{0.16V}{-S \ln (1 - \bar{a})}$$

Typical absorption coefficients

Absorption coefficient, a, is the fraction of the incident sound energy which is absorbed. For complete, 100 per cent absorption $a = 1$; for no absorption $a = 0$.

Material	Frequency Hz			
	63	500	1 k	4 k
225 mm brickwork	0.05	0.04	0.01	0.0
113 mm brickwork	0.10	0.05	0.0	0.0
75 mm breeze block	0.09	0.16	0.0	0.0
12 mm wood panels on 25 mm battens	0.33	0.33	0.10	0.12
Glass (> 6 mm thick)	0.03	0.03	0.03	0.03
3 mm hardboard on 25 mm battens	0.30	0.43	0.07	0.11
Brick (surface)	0.02	0.02	0.04	0.07
Rough concrete	0.01	0.02	0.06	0.10
Smooth concrete	0.01	0.02	0.02	0.05
Smooth plaster (painted)	0.01	0.01	0.02	0.02
Wood	0.05	0.07	0.10	0.12
Lino	0.02	0.02	0.03	0.04
Rubber flooring	0.01	0.03	0.04	0.02
Haircord carpet on underfelt	0.05	0.17	0.29	0.30
Wilton carpet on underfelt	0.04	0.22	0.64	0.71
Typical carpet tiles	0.01	0.11	0.39	0.55
Curtains, velour, draped	0.05	0.31	0.80	0.65
Lightweight fabric over 50 mm airspace	0.0	0.10	0.50	0.50

	Frequency Hz			
Material	**63**	**500**	**1 k**	**4 k**
25 mm mineral wool, 5% perforated hardboard cover	0.03	0.47	0.90	0.31
As above but with 25 mm airspace	0.04	0.65	0.90	0.31
As above but with 175 mm airspace	0.35	0.89	1.02	0.44
50 mm mineral wool, 5% perforated cover	0.10	1.10	0.90	0.31
50 mm mineral wool over 150 mm airspace	0.60	0.95	0.81	0.85
Audience (units/person)	0.15	0.40	0.45	0.45
Orchestra (units/person including instruments)	0.20	0.85	1.39	1.20

Typical recommended reverberation times

Activity	**T_{60} (s)**	**Comments**
Speech (1)	0.6–1.2	e.g. lecture theatres, Council chambers Conference halls
Speech (2)	1.0–1.4	Theatres
Reproduced sound	0.8–1.2	Cinemas
Multi-purpose use	1.0–1.5	School halls, Multi-function halls, Community halls

Recommended reverberation times for broadcasting

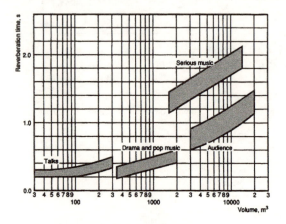

Figure 14 Sound studio reverberation times (BBC)

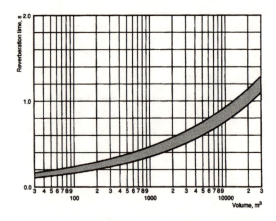

Figure 15 Television studio reverberation times (BBC)

9 Sound isolation

Noise design criteria

Single figure values in dBA cannot be taken as being reliable for design purposes as they give no indication of the frequency spectrum. The list below may nevertheless be of some use in stating an approximate indication of loudness (dBA) and also the approximate NC and NR (see Figures 16 and 17).

Environment	NC/NR index	Equivalent dBA
Entertainment:		
Concert hall, theatre	20–25	30–35
Lecture theatre, cinema	25–30	35–40
Hospital:		
Operating theatre	30–35	40–45
Multi-bed ward	35	45
Hotel:		
Individual room, suite	20–30	30–40
Banquet room	30–35	40–45
Corridor	35–40	45–50
Retail, restaurants:		
Restaurant, department store	35–40	45–50
Public house, cafeteria	40–45	50–55
Retail store	40–45	50–55
Industry:		
Light engineering workshop	45–55	55–65
Heavy engineering workshop	50–65	60–75

Environment	NC/NR index	Equivalent dBA
Domestic:		
Private dwelling, bedroom	25	35
Private dwelling, living room	30	40
Offices:		
Open plan office	35	45
Drawing office	35–45	45–55
Boardroom	25–30	35–40
Executive office	30–35	40–45
Public buildings:		
Court room	25–30	35–40
Library, bank, museum	30–35	40–45
Assembly hall	25–35	35–45
Sports arena	40–50	50–60
Church	25–30	35–40
Educational:		
Classroom, lecture theatre	25–35	35–45
Laboratory	35–40	45–50

Noise rating curves

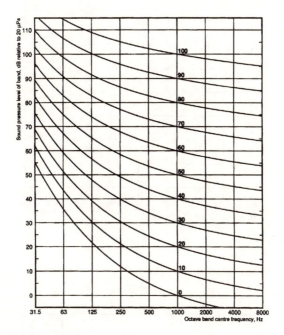

Figure 16 ISO noise rating (NR) curves

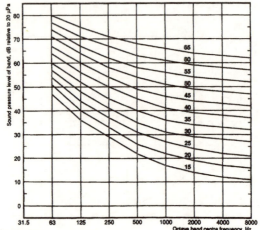

Figure 17 Noise criteria (NC) curves

Broadcasting studio criteria

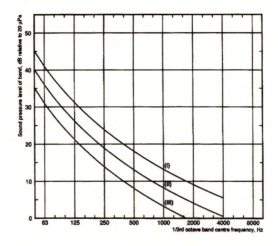

Figure 18 Background noise criteria for BBC studios
(i) relaxed criterion; (ii) normal criterion; (iii) radio drama studios

Airborne sound isolation

Formula for transmission loss introduced by a partition or barrier of surface mass m/m^2 at frequency f:

$$\frac{p_r}{p_s} = \frac{\rho_0 c}{\pi f m}$$

where p_r and p_s are the sound pressure levels on the receive and send sides of the barrier and $\rho_0 c$ is the characteristic impedance of the air.

Alternatively the Sound Reduction Index R is given by

$$R = 20 \log (fm) - 47 \text{ dB}$$

Effect of reverberation time on SRI

In the case of sound transmission from one room to another adjoining one the effective SRI, R, is affected by the reverberation time in the receiving room. Then

$$R = P_s + P_r + 10 \log \frac{ST}{0.163 \, V}$$

where P_s and P_r are the sound pressure levels in the source room and the receiving room respectively, S is the area of the partition dividing the two rooms, T and V are the reverberation time and volume of the receiving room.

Sound insulation performance for building materials (100–3kHz average)

The figures given in the following tables (pages 50–52) assume source and receiver are both rooms. If either is in the open air deduct 5 dB from the given value as an approximate correction.

Material	Approx. sound level reduction (dB)
WALLS	
Light unplastered blockwork	35
Light plastered blockwork	40
Solid 100 mm unplastered brick	42
Solid 100 mm brick with 12 mm plaster	45
Sealed dense concrete, 150 mm	47
Solid unplastered brickwork, 230 mm	48
Solid plastered brickwork, 230 mm	49
Dense concrete, sealed and plastered, 200 mm	50
Plastered brick, 450 m	55
225 mm double-sided brick wall with 50 mm cavity	65
TIMBER-FRAMED STRUCTURES	
50 mm frame with 12 mm plasterboard on each side	33
50 mm frame filled with quilt and 2×12 mm plasterboard each side	41
As above but 75 mm frame	45
FLOORS	
21 mm boards or 19 mm chipboard	35
110 mm concrete and screed	42
21 mm boards or 19 mm chipboard on plasterboard and 50 mm sand pugging	45
200 mm reinforced concrete and 50 mm screed	47
150 mm concrete on specialist floating raft	55–60

Material	Approx. sound level reduction (dB)
DOORS	
Hollow core, well-fitted panelled door, no seals	15
As above with seals	20
Solid core, well-fitted door with no seals	15
As above, with seals or cut close to carpet	25
Solid core door, 60+ mm with good seals	30

Typical sound reduction indices

Material		Octave band centre frequency (Hz)				
	kg/m^2	125	250	500	1 k	2 k
4 mm glass	10	20	22	28	34	34
6 mm glass	15	18	25	31	36	30
6.4 mm glass laminated		22	24	30	36	33
12 mm glass	30	26	30	35	34	39
Double glazed, sealed units: glass/airspace/glass						
3/12/3		21	20	22	29	35
6/12/6		20	19	29	38	36
6/12/10		26	26	34	40	39
6/20/12		26	34	40	42	40
Single leaf brick, plastered both sides	240	34	37	41	51	58
Cavity brickwork with ties	480	34	34	40	56	73
Double leaf brickwork plastered both sides	480	41	45	48	56	58

Material		Octave band centre frequency (Hz)				
	kg/m²	125	250	500	1 k	2 k
Three leaf brickwork plastered both sides	720	44	43	49	57	66
9 mm plasterboard on 50 × 100 mm studs at 400 mm centres		15	31	35	37	45
13 mm plasterboard on studs as above		25	32	34	47	39
13 mm plasterboard as above with 25 mm mineral wool between studs		25	37	42	49	46
9 mm ply on frame	5	7	13	19	25	19
25 mm T&G timber boards	14	21	17	22	24	30
Two layers 13 mm plasterboard	22	24	29	31	32	30
6 mm steel sheet	50	27	35	41	39	39
43 mm flush, hollow-core door, normal hanging	9	12	13	14	16	18
43 mm solid core door, normal hanging	28	17	21	26	29	31
50 mm steel door with good seals		21	27	32	34	36

10   Microphones

Basic requirements

Before purchasing microphones for high quality applications, such as recording or broadcasting, the following check-list should be considered.

Sound quality

1. Is the frequency response satisfactory? Ideally this should be $\pm 2\,dB$ over the range 20 Hz to 20 kHz but less tight tolerances may be acceptable at the high and low frequency ends. There can be advantages in a non-flat response for some specialized purposes: a reduced bass response can be an advantage in public address situations, for example.

2. Transient response. This is not easy to assess except by listening tests, which should be done with a wide range of sound sources.

3. Sensitivity. Is the electrical output adequate bearing in mind the situations in which the microphone will be used?

4. What is the self-generated noise? This is sometimes quoted as the equivalent acoustic noise in dBA. For digital recording work the microphone's noise level should obviously be as low as possible.

5. What is the maximum sound pressure level which can be handled without distortion? (Note that the limit is generally set at the first stage of amplification, which in the case of electrostatic microphones is inside the microphone itself.)

6. How much variation is there between different samples of microphone of the same type? This may be considerable with cheap devices.

Directional response

7. What is the directivity pattern (polar diagram)?

8. How much does this vary with frequency and does it depend on the orientation of the microphone: i.e. is it different in the vertical plane compared with the horizontal plane?

9. Is the directivity fixed or variable? If variable is the control on the body of the microphone or is there a remote unit? What patterns are available and how good are they?

Physical

10. Physical dimensions.

11. Weight.

12. Appearance, including colour and finish: e.g. matt or shiny.

Reliability

13. Can trouble-free service be expected? This can probably only be ascertained by referring to other users with experience of this microphone.

14. Does it appear to be robust?

15. Are there any reasons for thinking that there might be deterioration with age and use?

16. How likely is it that on-site maintenance can be carried out, how available are spares, what repair service, including turn-round time can the manufacturers offer?

Vulnerability

17. Is it prone to wind noise, 'popping', etc?

18. Is there a windshield available? If so how effective is it (are they)?

19. To what extent is the microphone likely to be affected by humidity and temperature?

20. Is it reasonably immune to external magnetic and electrostatic fields, including r.f. pick-up?

21. How prone is it to the effects of vibration, handling noise, cable rustle, etc.?

Electrical

22. What is the optimum electrical load?

23. What power supplies does it need, if any?

24. If capable of being battery-operated what is the battery life and how readily available are the batteries?

25. Connectors. Are these standard?

26. Is there built-in frequency correction, e.g. a bass-cut switch?*

27. Is there a built-in, switchable, attenuator?*

(* The absence of these facilities does not necessarily imply an inferior microphone as it may well be that there is never likely to be a need for them.)

Miscellaneous

28. What types of mounting are available?

29. What is the cost?

30. What is the reputation of the manufacturer?

31. If made abroad are there good agents in this country?

Microphone sensitivities in commonly-used units
See table on following page.

Microphone transducers

Moving coil
The sensitivity of the average moving coil microphone is such that an e.m.f. of the order of 0.5 to 1 mV is produced with normal speech at a distance of 0.5 m.

Microphone sensitivities in commonly-used units

dB relative to 1 V/Pa	mV/μbar	mV/10μbar	Comments
− 20	9.5	95	Typical electrostatic microphones
− 25	5.5	55	
− 30	3.0	30	
− 35	1.8	18	
− 40	1.0	10	
− 45	0.55	5.5	
− 50	0.30	3.0	Typical moving coil
− 55	0.18	1.8	
− 60	0.10	1.0	Some ribbon mics.
− 65	0.055	0.55	

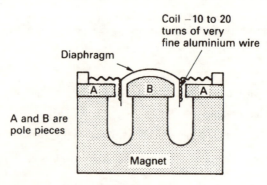

Figure 19 Moving coil microphone (simplified)

Typical characteristics:

1. Usual directivity patterns are cardioid, omnidirectional or 'gun'. Figure-of-eight responses need two back-to-back units and have rarely been satisfactory.

2. Reasonably immune to humidity and temperature.

3. Moderately robust.

4. They have the advantage of not needing power supplies.

Ribbon

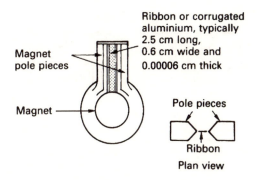

Figure 20 Basic ribbon microphone

The ribbon impedance is usually of the order of 1 Ω and a built-in transformer is used to raise the impedance at the output to a much higher value. This has the added benefit of raising the output voltage but even so this is small, being less than 1 mV for normal speech at a distance of 0.5 m.

Typical characteristics:

1. The most usual directivity patterns are figure-of-eight or hypercardioid.

2. Fragile. The ribbon needs to have effective screening from air blasts inside the microphone casing. They are therefore not suitable for use out of doors.

3. Sensitivity is almost always low.

4. A specialized use is in the 'lip ribbon' microphone used for commentaries in noisy environments (see later).

The combination of low sensitivity, relative fragility and limited range of directivity patterns means that ribbon microphones tend to be used rarely, the lip ribbon being something of an exception.

Electrostatic

In simple form the basic circuit for an electrostatic (sometimes called 'condenser' or 'capacitor') microphone is shown in Figure 21.

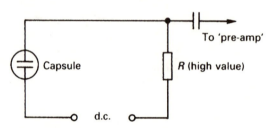

Figure 21 Essential circuit for an electrostatic microphone

The spacing between diaphragm and back plate is typically about 0.02 mm, forming a capacitor of value about 20 pF. The d.c. supply may be in the range 50 V to 100 V, with R having a very high value of several hundred MΩ. The effect is to give a relatively long time constant to the CR combination. Then, with a variation of δC in the capacitance there is a change of the voltage given by

$$\delta V = Q \, \delta C$$

Electrets – materials with a permanent electrostatic charge are often used either for the back plate or the diaphragm and this removes the need for a polarizing voltage.

r.f. electrostatic – the capsule forms part of an LC circuit which in turn is part of an r.f. discriminator. The output of an oscillator, typically about 8 MHz, is fed into the discriminator. The output of the latter is an audio signal representing the variations of the capsule's capacitance.

The advantage of this type of transducer is that it is largely unaffected by humidity, although the cost is likely to be high (see Figure 22).

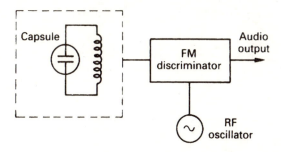

Figure 22 Simplified r.f. electrostatic microphone

Typical characteristics of electrostatic microphones.

1. Good frequency response because of the lightness of the diaphragm. The high frequency response is generally very good because the diaphragm is under tension, resulting in a high resonant frequency.

2. High sensitivity.

3. All types of directivity pattern (polar diagram) can be produced.

4. Prone to humidity problems, although careful drying in a warm environment can generally restore normal operation within 30 to 60 minutes.

Directivity Patterns (Polar Diagrams)

Omnidirectional
Essentially the result of pressure operation on the diaphragm – i.e. the force on the diaphragm depends on the sound wave pressure only and not a derivative of it.

'Omni' microphones are only truly omnidirectional when the incident sound wavelengths are large compared with the microphone diameter.

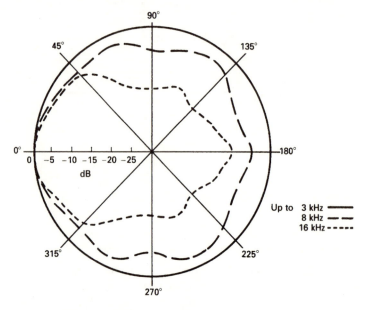

Figure 23 Directivity patterns for a typical 'omni' microphone

A useful approximate calculation to find the frequency up to which the microphone is omnidirectional results from finding the frequency corresponding to the diameter of the microphone. The microphone will be reasonably omnidirectional up to frequencies between two and three octaves below that.

For example, a microphone has a diameter of 2 cm. The frequency corresponding to 2 cm wavelength is 17 kHz. One octave below this is about 8 kHz; two octaves, 4 kHz; three octaves, 2 kHz. Thus this microphone will be reasonably omnidirectional up to about 3 kHz.

Phase cancellation, i.e. a reduction in sensitivity to sounds from the side, may also occur when there is a complete wavelength, or approximately so, across the diaphragm, as shown in Figure 24.

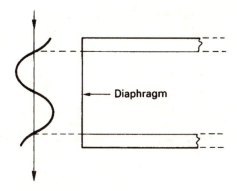

Figure 24 Phase cancellation

Figure 25 shows the frequency response at different angles of sound incidence for a typical omni microphone.

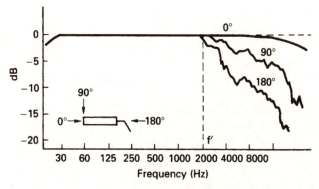

Figure 25 Typical omni microphone frequency response graphs

General characteristics of omni microphones are:

1. Usually less affected by rumble and handling noise than other types.

2. Do not exhibit proximity effects (see under figure-of-eight microphones).

3. While not offering any advantages in rejection of sounds from certain angles their relatively constant sensitivity, except at high frequencies, can sometimes be useful.

Figure-of-eight microphones

The force on the diaphragm results from the pressure gradient (pressure difference) between its two sides.

A figure-of-eight pattern may be represented by

$$r = \cos \theta$$

where r is the effective sensitivity at angle θ.

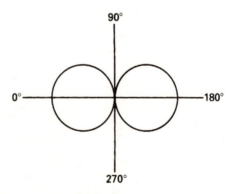

Figure 26 Figure-of-eight polar diagram

The pressure gradient may be calculated from

$$F_G^2 = F_f^2 + F_t^2 - 2F_f F_t \cos \beta$$

where F_G is the pressure gradient, F_f and F_r are the forces on the front and rear of the diaphragm respectively, and β is the angle found from

$$\beta = \left(\frac{df}{c} \right) \times 360$$

d being the acoustic distance between the front and rear of the diaphragm, f the frequency and c the velocity of sound.

Figure 27 illustrates the above.

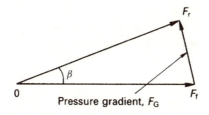

Figure 27 Method of determining the pressure gradient

If the distance of the sound source is large compared with d so that inverse-square-law effects can be neglected the expression for F_G simplifies to

$$F_G^2 = 2F_f^2(1 - \cos \beta)$$

The variation in pressure gradient with frequency is illustrated in Figure 28.

Proximity effect ('Bass tip-up')

This is shown when the source of sound is close to the microphone so that, because of inverse-square-law effects, F_f and F_r are not equal and the frequency f is low so that β is small. The result is that the output of the microphone for low frequency sounds becomes exaggerated (see Figure 29).

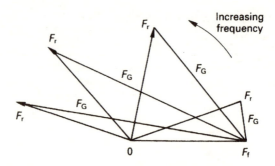

Figure 28 The variation in pressure gradient with frequency

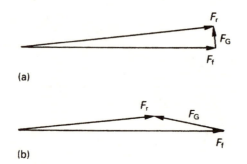

(a)

(b)

Figure 29 Proximity effect

The effect is used to good advantage in certain microphones used by radio or out-of-vision television commentators – the so-called 'lip ribbon' microphone. Bass cut restores the voice to an approximately normal response but reduces the level of distant l.f. noises.

The table below gives the increase in microphone output, relative to the output at 10 kHz for a typical microphone of this type.

General characteristics of figure-of-eight microphones:

1. Very prone to rumble and vibration.

Proximity effect for a pressure gradient microphone (for a point source, 50 mm from daphragm)

Frequency (Hz)	Relative output (dB)
50	26
100	20
200	14
500	7
1 k	3
2 k	1
10 k	0

2. Show proximity effect (see above).

3. The figure-of-eight pattern is generally fairly well maintained with frequency although the nulls at 90° to the axis may be poorly defined at some frequencies.

Cardioid

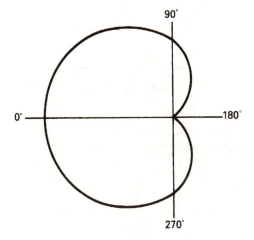

Figure 30 Cardioid pattern

The cardioid pattern in Figure 30 is a plot of

$$r = 1 + \cos \theta$$

In practice this is rarely obtained with a 'cardioid' microphone. The frequency response graph of a typical high-grade unit is shown in Figure 31 from which it can be seen that the 180° response, far from being zero, may be no more than about 10 dB below the axial response at low frequencies, and possibly even worse at high frequencies.

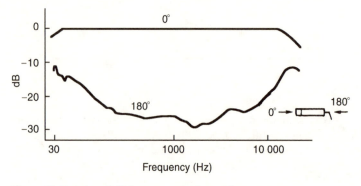

Figure 31 Typical cardioid frequency response graphs

General characteristics of cardioid microphones:

1. Usually show some proximity effect.

2. Tend to be prone to rumble and vibration effects.

Hypercardioid

The term 'hypercardioid' is often used for any microphone with 'dead' angles in the rear quadrants, but the most common usage is for dead angles at 45° off the 180° axis. (See Figure 32.) The equation in that case is

$$r = \frac{1}{\sqrt{2}} + \cos \theta$$

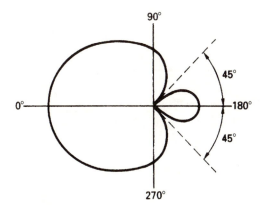

Figure 32 Hypercardioid pattern

Hypercardioid microphones can be thought of as being half way between figure-of-eight and cardioid microphones, and their general characteristics reflect this.

General characteristics of hypercardioid microphones:

1. Proximity effect is typically less than for cardioids but more than for figure-of-eights.

2. Rather prone to rumble effects.

Variable directivity microphones
The capsule of a variable-directivity microphone normally consists of two back-to-back electrostatic cardioid microphones, the potential on the diaphragm of the rear unit being variable as shown in Figure 33.

Highly directional microphones
Interference-tube ('Gun') microphones (see Figure 34).

Typical polar diagrams are given in Figure 35.

Interference tube microphones of about 250 mm length often have polar patterns which are close to hypercardioid in shape.

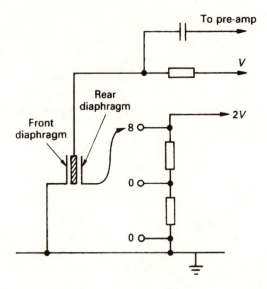

Figure 33 Simplified circuitry for a variable directivity pattern microphone

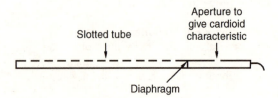

Figure 34 Simplified interference tube

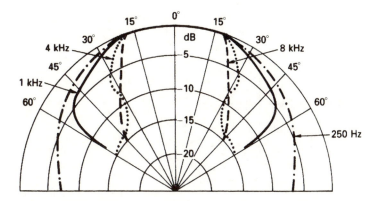

Figure 35 Typical directivity plots for an interference tube microphone of 500 mm length (approx.)

Phantom power systems

1. 48 V standard systems (sometimes indicated by a 'P' in the type number).

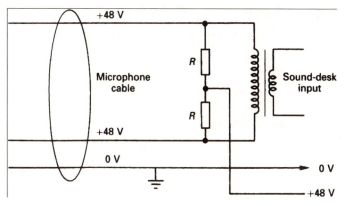

Figure 36 48 V phantom power – the supply end

The resistors R, in Figure 36, are for current-limiting in the event of a short circuit. A common value is 6.8 kΩ.

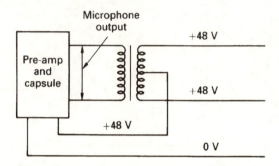

Figure 37 48 V phantom power – microphone end

2. A–B powering

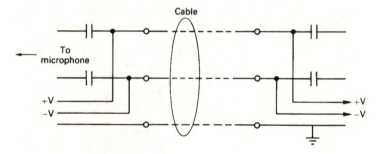

Figure 38 A–B powering

A–B powering is generally a low-voltage method – typically 7–9 V – used with single microphones, whereas the 48 V system is usually applied to all the microphones in a studio. It has the advantage that only two wires are needed. The disadvantage is that a phase reversal will result in the failure of the microphone to work until the reversal is corrected.

11 Radio microphone frequencies

VHF frequencies

News frequencies	Shared frequencies	Coordinated frequencies	Deregulated frequencies
		173.800	173.800
		174.100	174.100
		174.500	174.500
		174.800*	174.800*
		175.000	175.000
	175.250		
	175.525	176.400	
176.800	176.600	177.000	
184.600			
184.800			
185.000			
191.700	191.900	192.300	
192.100	192.800		
192.600	193.000		
199.900	199.700		
	200.300	200.100	
200.800	200.600		
201.000			
207.900	208.300	207.700	
208.800	208.600	208.100	
	209.000		
216.300	216.100		
217.000	216.600		
	216.800		

*This frequency is not fully compatible with the other Coordinated/Deregulated frequencies and may result in intermodulation.

The information in this chapter is reproduced by courtesy of John Willets, MIBS, of Sennheiser UK and the Institute of Broadcast Sound. Radio microphone frequencies tend to be changed from time to time and the reader is advised to check with JFMG (the licensing authority) whose address is given at the end of this section.

In the table shown on page 72 News frequencies are what the term implies – they may be used by anyone engaged in news gathering. Shared frequencies are for general use anywhere in the UK but before using a radio microphone checks should be made that there is no one in the vicinity using the same frequencies. Coordinated frequencies are for use at predetermined sites and must be cleared with JFMG beforehand. Deregulated frequencies can be used by anyone with type-approved equipment.

Shared frequencies: TV Channel 69:

Frequency	Notes
854.900	For general use indoors or outdoors
855.275	
855.900	
860.400	
860.900	Users should beware of causing interference to adjacent TV services
861.750	
856.175	
856.575	
857.625	
857.950*	
858.200	
858.650	
861.200	
861.550*	

*These frequencies, although legal, may cause problems with some equipment due to their closeness to adjacent channels.

Coordinated frequencies for the London area

This table lists the channels allocated to radio microphones only. The full allocation list is available from JFMG, as are similar lists for many other areas of the country.

TV Channel	Notes
36	Shared with Radar
38	Shared with Radio Astronomy
39	May be available with channel 38
42	3rd choice for new assignments
43	3rd choice for new assignments
44	3rd choice for new assignments
45	3rd choice for new assignments
46	3rd choice for new assignments
47	3rd choice for new assignments
48	3rd choice for new assignments
56	2nd choice for new assignments
57	2nd choice for new assignments
58	2nd choice for new assignments
59	Satisfactory for continued use
60	Satisfactory for continued use
61	Satisfactory for continued use
66	1st choice for new assignments
67	1st choice for new assignments
68	1st choice for new assignments

Licensing

It is a requirement that a transmitter is licensed before use and the frequency is cleared with

JFMG Limited
72 Upper Ground
London SW1 9LT
Tel: 020 7261 3797
Fax: 020 7737 8499
Email: jfmg@compuserve.com
Website: www.jfmg.co.uk

12   Loudspeakers

Moving-coil drive units

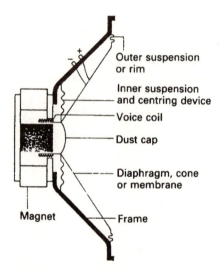

Figure 39 Basic elements of a moving-coil unit

The magnitude of the force on the coil is given by

$$F = BIl \text{ newtons}$$

where B is the flux density (Wb/m²), I is the current (A) and l is the coil length (m).

The velocity of the coil/cone is found from

$$v = F/Z \text{ m/s}$$

where F is the force (N) and Z is the mechanical impedance (mech. Ω).

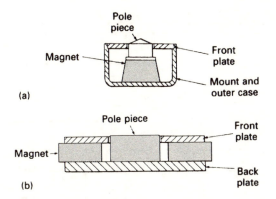

Figure 40 Basic magnet configurations: **(a)** conventional central 'pot' magnet; **(b)** flat shape used with ceramic magnets

To avoid distortion with large amplitude signals, and hence large coil movements, it is necessary for the length of the coil effectively within the magnetic field to be constant. Figure 41 shows typical arrangements.

Peak excursion v frequency. Figure 42 shows the relationship for different cone diameters radiating 1 W.

Ribbon loudspeaker
Similar in basic design to the ribbon microphone (q.v.). It is difficult to make them efficient at low frequencies, so they are mostly used as tweeters.

Piezo-electric loudspeaker
Again, used for high-frequency units, these make use of the property of certain materials to undergo physical deformation when a voltage is applied.

Multiple driver units
Because it is almost impossible to cover the entire audio range satisfactorily with one drive unit it is common practice to use two or more units, each designed to cover a particular range of frequencies. 'Crossover units' act as band-pass filters to ensure that each unit is fed only with the appropriate frequencies.

(Continued on page 80)

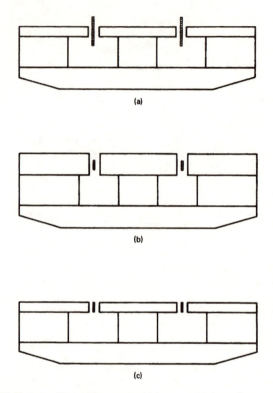

Figure 41 Typical coil/gap arrangements. (c) is acceptable in small excursion systems such as mid-range units and tweeters

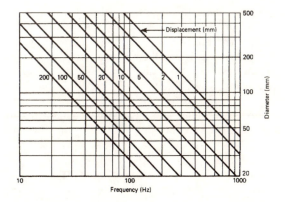

Figure 42 Peak coil excursions against frequency for different coil diameters to radiate I W

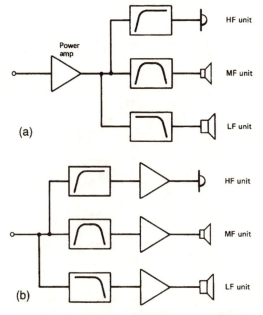

Figure 43 Crossover networks: **(a)** high-level, or passive; **(b)** low-level, or active

(Multiple driver units, continued from page 77)

Broadly, there are two ways of powering such systems: 'high-level', or passive, in which there is one power amplifier but each filter circuit has to handle a high-power signal, Figure 43(a); 'low-level', or active, where the filters operate at low level and there is an amplifier for each drive unit, Figure 43(b). This has the advantage that relatively small and inexpensive components are used in the filters and this can outweigh the cost of the extra amplifiers.

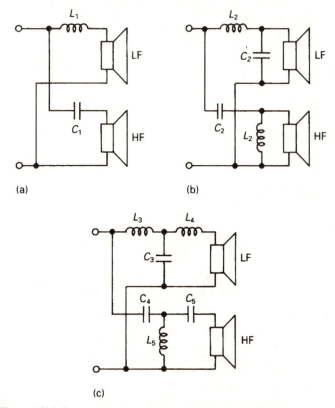

(a)

(b)

(c)

Figure 44(a) First-order crossover network giving a slope of 6 dB/octave;
(b) Second-order crossover network giving a slope of 12 dB/octave;
(c) Third-order crossover network giving a slope of 18 dB/octave

Types of crossover network

Figure 44 (a), (b) and (c) (page 80) show the basic circuits for networks giving roll-off slopes of 6, 12 and 18 dB/octave. Corresponding response curves are given in Figure 45.

(Note that the circuits shown in Figures 44 (a), (b) and (c) assume that the nominal impedence is the same for all drivers and is constant at all frequencies – i.e. resistive.)

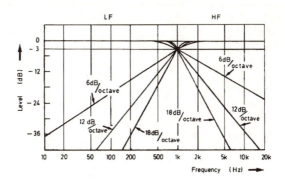

Figure 45 Response curves for first-, second- and third-order crossover networks

Baffles and Enclosures

1. Flat baffle

Cheap and simple but poor bass response. The lower frequency limit is given approximately by $L = 2\lambda$ where L is the shortest baffle dimension and λ is wavelength. For a square baffle of side 1 m this gives a lower frequency limit of roughly 700 Hz.

2. Closed box (sometimes termed 'infinite baffle')

The bass response is not as good as might be expected because the enclosed air acts as a spring, the stiffness of which raises the bass resonance frequency of the system. Nevertheless, with careful design very effective small loudspeakers are produced with sealed box enclosures.

3. Vented enclosure ('bass reflex')

The entire enclosure contains a vent, so forming a Helmholtz resonator which is tuned so that the air in the vent is in phase with the cone of the main drive unit.

Instead of a vent auxiliary bass resonator (ABR) is sometimes used. This is a passive device, either flat or cone-shaped.

4. Transmission line ('labyrinth')

Back radiation from the cone travels down a long pipe lined with sound absorbent material. Its design should be such that sound emerges at the end in phase with the radiation from the cone. Theoretically the line should be a quarter of a wavelength long at the bass resonance frequency, i.e., for lower frequency limit of 30 Hz the line should be about 3 m long. Careful damping and tapering are needed to avoid the effects of standing waves.

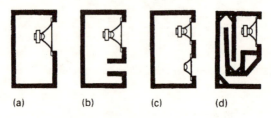

(a)   (b)   (c)   (d)

Figure 46 Loudspeaker enclosures. **(a)** closed box; **(b)** vented enclosure; **(c)** vented enclosure using ABR; **(d)** transmission line

5. Horn loading

A suitably flared horn can increase greatly the efficiency of a loudspeaker – up to 5 per cent compared with possibly as little as 1 to 2 per cent obtainable with other systems. Dimensions and rate of flare are critical and the horn itself should not vibrate (as can happen with the metal horns often found at, say, race tracks). For a good bass response the horn should be several metres long.

Loudspeaker impedance and frequency

This varies widely with frequency, as shown in Figure 47. The quoted impedance, often 8 Ω, is usually the value measured at 400 Hz. This is often about 20 per cent higher than the d.c. resistance of the coil.

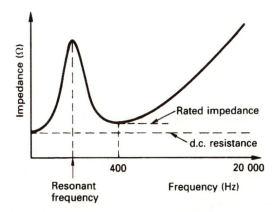

Figure 47 Impedance versus frequency for a typical closed box loudspeaker

The most-often quoted, and frequently the most useful indication of a loudspeaker's sensitivity is the sound pressure level (SPL) at 1 m distance in free-field conditions for a given input signal level. The latter may be 2.83 V, corresponding to 1 W into a nominal impedance of 8 Ω. Alternatively the peak SPL at 1 m before significant distortion occurs may be given. For professional monitoring loudspeakers 120 dBA is a typical rating.

13 Stereo

Time-of-arrival (TOA) difference

This is the difference in arrival time of a sound at the two ears of the listener. It is probably the most important single factor in determining the direction of a sound. It is zero for sounds arriving from directly in front of the listener and is a maximum of about 0.6 ms (depending on the size of the head) for sounds arriving at 90° to the frontal axis. In general the TOA difference is proportional to sin θ in Figure 48.

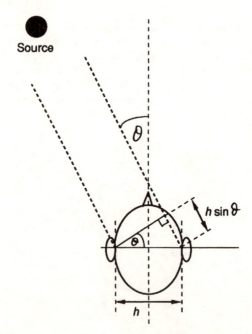

Figure 48 Time of arrival difference, t is given by $t = h \sin \theta / c$ where c is the velocity of sound in air

Pure tones are generally very difficult, or even impossible, to locate and there needs to be a low frequency component. This may take the form of a modulation of an otherwise steady tone.

The Haas effect

If the same sound is emitted by two sources, as would occur for example if a mono signal were fed into two similar loudspeakers, the brain appears to fuse together the two signals provided that the time difference between them is less than about 50 ms. There is a perceptible time interval if the difference is greater than about 50 ms.

With differences less than 50 ms not only do the two sources appear as one but the apparent single source tends to be towards origination of the first arrival. For the delayed source to seem as loud as the undelayed one it must be higher in level by an amount which depends on the time delay. The relationship is shown in Figure 49.

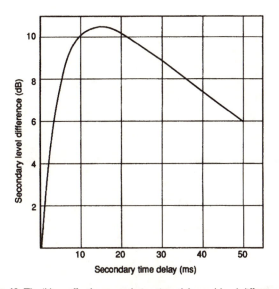

Figure 49 The 'Haas effect' curve relating time delay and level difference for two sound sources to appear to have the same loudness

Sound image positions with two loudspeakers

TOA differences, while very important in real-life location of sounds, do not apply in the case of stereo reproduction from two loudspeakers. It can be shown that sound image position depends upon the 'inter-channel difference', i.e. the number of dBs difference in level between the left and right signals. The relationship is shown in Figure 50.

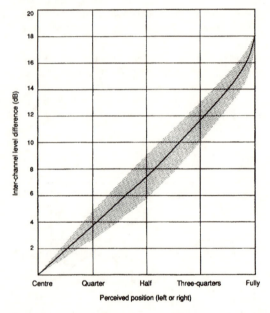

Figure 50 Relationship between interchannel difference and image position

Figure 50 was originally produced from speech sources but it appears to be applicable to other kinds of sound. The shaded area indicates the disagreement which can occur between different listeners. Interestingly, if the signals to the loudspeakers are changed over, so that what was the left signal is fed to the right speaker, and the right signal is fed to the left speaker, for many listeners the shape of the curve is different. That is, it is not a mirror image of the original.

Stereo terminology

Application	Channel	
	Left	**Right**
Designation in domestic equipment	L	R
Colour in domestic equipment (including headphones)	White (or no colour)	Red
Designation in broadcast equipment	A	B
Designation in some stereo microphones	X	Y
Colour in broadcast equipment*	Red	Green

* Note that the broadcasting colours follow the colour convention used for navigation lights at sea.

M and S signals

The M signal is essentially the sum of the signals in the two channels; the S signal is the difference:

$$M = A + B \qquad S = A - B$$

In practice a correction is applied. The addition of A and B, if these were identical as is the case for a central stereo image, then M is 6 dB higher than A or B. The correction which is used depends on circumstances, and sometimes on the broadcasting organization – it may be 3 dB or 6 dB.

Hence

$$M = A + B - 3\,dB \qquad or \qquad M = A + B - 6\,dB$$

Similarly

$$S = A - B - 3\,dB \qquad or \qquad S = A - B - 6\,dB$$

Derivation of M/S signals from A/B and vice versa

Two methods of doing this are shown in Figure 51.

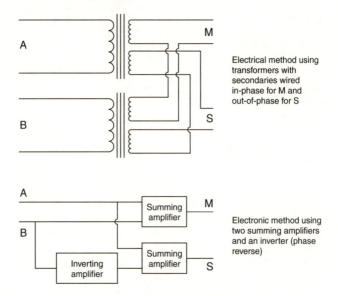

Electrical method using transformers with secondaries wired in-phase for M and out-of-phase for S

Electronic method using two summing amplifiers and an inverter (phase reverse)

Figure 51 Conversion of stereo signals from A/B to M/S (or M/S to A/B).

Microphones for Stereo

1 Coincident pair microphones, A/B (or XY) configuration

Both microphones have, as near as is practicable, identically similar characteristics and their diaphragms are mounted as close together as possible (ideally they would be coincident in space).

As Figures 52 and 53 show, for two figure-of-eight microphones the conversion of an A/B system to an M/S one can be effected by turning the microphone through 45°! The out-of-phase regions may or may be significant. Individual sound sources in these areas will appear in the stereo image as difficult to locate. Reverberation, in for example, a concert hall, is sufficiently diffuse for there to be no apparent ill-effects for the components of the reverberation which enter the microphones from the out-of-phase angles.

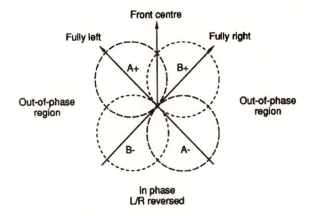

Figure 52 Polar diagrams for a pair of coincident figure-of-eight microphones. Note the out-of-phase quadrants

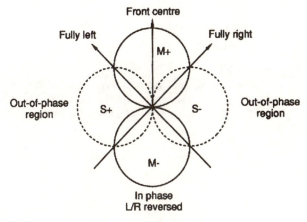

Figure 53 The equivalent M/S diagrams to Figure 52

Approximate stereo image positions

The following tables assume well-matched coincident-pair microphones angled at 90° to each other. The positions of the sound images assume well-matched loudspeakers and good listening conditions. They can only be approximate because listeners can differ in their placing of such images (see Figure 50). The angle of sound incidence is with reference to the centre line between the microphones.

1. Figure-of-eight microphones

Angle of sound incidence (°)	Approximate image position
− 45	Fully left
− 30	3/4 L
− 15	1/2 L
0	Centre
15	1/2 R
30	3/4 R
45	Fully right

The level of the M signal is about − 3 dB at 45° compared with the level at 0°.

2. Hypercardioids
Note that the polar diagrams for most hypercardioid microphones vary considerably with frequency. The figures in the table below are, nevertheless, likely to be fair indications for most speech and music.

Angle of sound incidence (°)	Approximate image position
− 75	Fully left
− 60	3/4 L
− 45	1/2 L
− 20	1/4 L
0	Centre
20	1/4 R
45	1/2 R
60	3/4 R
75	Fully right

The level of the M signal is about –3 dB at roughly 60° to 70° compared with the level at 0°.

3. Cardioids

Note that, like hypercardioids, the polar diagrams for most cardioid microphones vary considerably with frequency. The figures in the table are, again, likely to be fair indications for most speech and music.

Angle of sound incidence (°)	Approximate image position
− 90	Fully left
− 60	1/2 L
− 30	1/4 L
0	Centre
30	1/4 R
60	1/2 R
90	Fully right

The level of the M signal is about −3 dB at roughly 75° compared with the level of 0°.

4. M/S microphones

Basically any M/S microphone must consist of a sideways figure-of-eight microphone with a forward-facing microphone which can, in principle, be of any pattern. A commonly used system consists of a forward-facing cardioid with the inevitable sideways figure-of-eight (Figure 54).

An M/S signal must, sooner or later, be converted into an A/B signal, certainly for aural monitoring. Methods for doing this are illustrated in Figure 55.

The width of the stereo image is dependent on the amplitude of the S signal – if that is zero then there is no stereo component and the result is mono. A method of obtaining an A/B signal from M/S using three channels on a mixer is shown in Figure 55, where a fader in the third channel controls the S signal and is thus a width control.

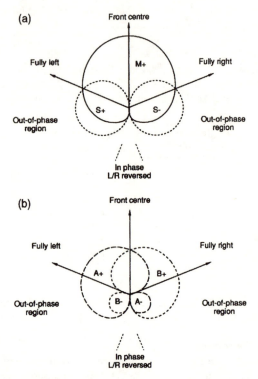

Figure 54 (a) M/S microphone, forward-facing cardioid; **(b)** the A/B equivalent – two cardioids at 90°

5. Panpot systems

Each microphone's output is placed in the stereo image by a 'panpot' which is essentially a potential divider which distributes the signal between the A and B channels according to the position of the sliding contact. It is, of course, possible to use coincident pair microphones and panpotted microphones at the same time.

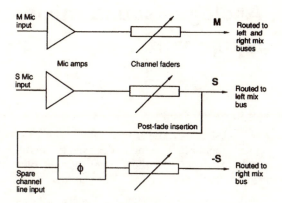

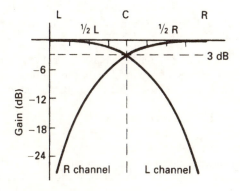

Figure 55 Derivation of A/B from M/S using three mixer channels, with the fader on the 'spare' channel providing width control

Figure 56 Panpot law

14 Analogue sound mixing equipment

Levels and dynamic range

1. Low level signals. These are generally accepted as the levels from microphones working under average conditions. This may be between -70 dBu (where the unit dBu takes as reference 'zero level' or 0.775 V) to around -50 dBu.

2. High level or 'line level' signals. Signals in the range from about -15 dBu to $+20$ dBu. The terms 'low level' and 'high level' are not intended to be precisely defined.

3. Dynamic range. The range in dB between the lowest and the highest programme levels. The lowest acoustic level likely to be encountered is about 20 dBA; the highest can be 110 dBA or may be more. The acoustic dynamic range is thus $90+$ dB.

   Digital recording systems such as compact discs can handle dynamic ranges of around 90 dB; f.m. radio, about 45 dB while a.m. radio is restricted to 30 dB or even less.

Balanced and unbalanced circuits

The difference is illustrated in Figure 57.

The important point about balanced circuits is that the two legs are electrically identically similar, so that any induced interference will produce, ideally, exactly equal voltages in each leg. These are in opposition and therefore will 'cancel out'. Perfect balancing is almost impossible, even if only because the two conductors inside the cable can never be at exactly the same distances from the sources of interference. Twisting the conductors improves matters and in one type of microphone cable ('star quad') four twisted conductors are used, opposite conductors being used for each leg. Such an arrangement gives very good immunity to the effects of interference, but even so 100 per cent protection can never be guaranteed.

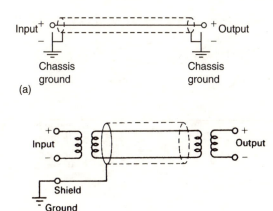

Figure 57 (a) unbalanced; **(b)** balanced circuits

Input impedances
Microphone: channel input impedances are normally about 1 kΩ.
  Channel: Input impedances are normally 10 kΩ.

Jack connecting plugs
The jacks most commonly used in audio work are 6.35 mm (¼ inch)
in diameter. There are two types in common use:

A-gauge. These can be either two-pole (unbalanced) or three-pole
(balanced). The latter can be used for balanced microphones or
unbalanced stereo. Their use is confined to domestic and non-
professional applications.

B-gauge (IEC 60268–12). These are used in professional applications, are always in the three-pole configuration and are made from
solid brass. Note that the tip and ring have smaller diameters than
the sleeve. They are thus not interchangeable with A-gauge jacks.

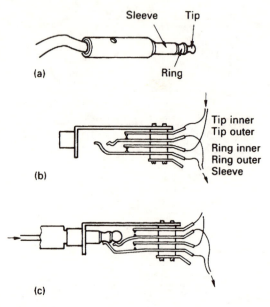

Figure 58 Jack plugs and sockets

Methods of inter-connection

1. 'Normalling'. Connections are semi-permanent but may need to be interrupted. 'Full normalling' requires jacks to be inserted in both sockets. 'half-normalling' needs only one jack inserted to break the circuit. See Figure 59.

2. A 'listen' jack allows a plug to be inserted without breaking the circuit.

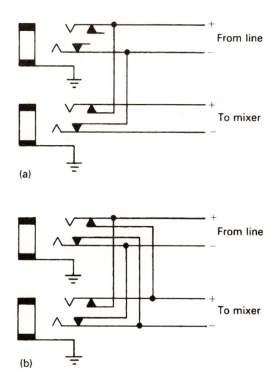

Figure 59 (a) Half-normalling; **(b)** full normalling

Channel facilities

Figure 60 shows a typical mono channel. This does not mean that the mixer is incapable of handling stereo signals, but that a mono channel could, for example be used for one half of a stereo source and two channels would thus be in effect one stereo channel. In such usage the panpots for the two channels would probably (but not necessarily inevitably) be set fully left and fully right.

The diagram is largely self-explanatory. However the following notes may be helpful.

1. The Dynamics unit is likely to be a limiter/compressor device with possible a noise gate on the more advanced mixers.

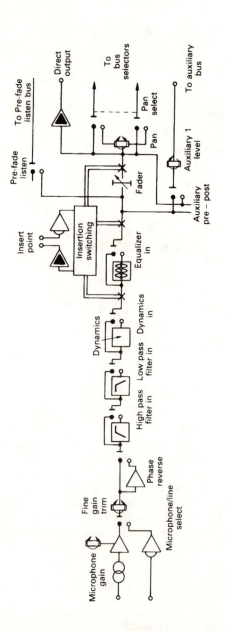

Figure 60 The facilities of a typical mono channel

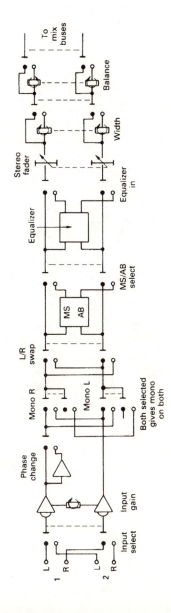

Figure 61 Typical stereo line channel. Auxiliaries, inserts, PFL etc. omitted for clarity

2. Additional 'outboard' devices can be plugged to the Insert point. A graphic equalizer is one such possibility.

3. Pre-fade listen (often designated PFL) allows an operator to monitor, either aurally or visually, or both, a source before it is faded up. This is particularly useful in live broadcasting where a contribution from a remote source needs to be checked before being faded up.

4. Auxiliary outputs could be used for a variety of purposes. While feeds to 'echo', PA (public address) etc. may be dedicated outputs on many mixers, on some installations where flexibility is needed auxiliary outputs can provide the same functions.

5. In Figure 60 one auxiliary output is provided with Pre-fader and Post-fader switching – i.e. the output can be derived either before or after the fader. In other words, in the Pre-fader mode alterations to the fader have no effect on the level of the auxiliary output; with Post-fader operation the auxiliary output level varies with fader setting. There can be advantages in having the choice of the two ways of working in some echo and PA situations.

6. Note that the filters, dynamics unit, etc. can be bypassed. This is desirable for several reasons – the unit can be switched out quickly if it develops a fault; in setting up the mixer the effect of the unit can often be best assessed by switching it in and out; having set the controls on the unit it may not be required until a later stage in the recording or broadcast, in which case it is switched out until needed.

Figure 61 shows a typical stereo channel for line input (e.g. for stereo tape machines, CDs or other stereo sources. The balance control gives a limited (around 5 dB) amount of left or right total shift.

Additional Terminology

AFL. After-Fade Listen. A feed available for monitoring by the operator and taken after the fader, as opposed to before the fader as in PFL. One use is to check on suspected distortion in a source.

Solo. All other channels are muted. 'Solo-in-place' leaves the channel's image in the stereo scene unchanged.

Clean feed ('Mix minus'). An output from a mixer which lacks one or more of the mixer input signals. The American term 'Mix minus' is more descriptive than 'Clean feed' which is the normal term in the UK. An example in broadcasting is in a programme being relayed to several countries: music and sound effects, etc. might be sent to other countries' broadcasting organizations without the local commentary/introductions. The programme material without the speech would be a clean feed.

Groups. The selective combining of two or more channel outputs into one group allows the group to have, for example, a group fader which is then, in effect, a sub-master fader. On a large mixer the group modules may provide similar facilities such as equalization, derivation of echo and PA as the channel modules.

Monitoring

1. **Aural monitoring** (e.g. by loudspeakers). The following are typical controls:

   Stereo/mono.

   Mono on both (useful for checking the balance between L and R speakers).

   Mono on one LS, usually the L speaker. (This can give a degree of checking of the compatibility of a stereo signal with what will be heard on mono equipment. On some installations there may be a small 'domestic quality' loudspeaker as a further aid to assessing compatibility.)

   Phase reverse – frequently in the feed to the L loudspeaker.

   Dim – a switch provided 12 dB or more of attenuation.

   Balance – adjustment of the relative gains of the feeds to the two loudspeakers.

   Volume control.

2. **Visual.** Typical instruments are

   VU ('Volume Unit') meters. These are essentially only voltmeters.

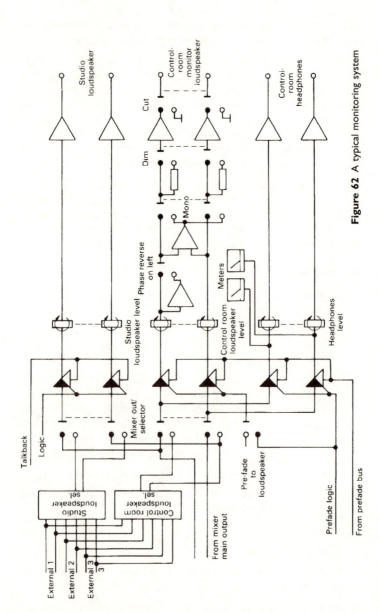

Figure 62 A typical monitoring system

PPM ('Peak Programme Meter'). Quasi peak-indicating instruments with a rapid pointer rise time and slow fall-back. The scale has white on black markings designed for ease of reading with minimal eye fatigue. The PPM is better than the VU meter for indicating incipient overloads. It is much more expensive. See also Section 20.

Stereo PPMs have concentric pointer spindles. Two stereo PPMs are often mounted side-by-side, and then the pointers are colour-coded as follows:

PPM 1: white = M signal and equivalent to a mono PPM
yellow = S signal

PPM 2: red = L (A) signal
green = R (B) signal

A typical monitoring system is shown in Figure 62.

15 Signal processing

Basic 'tone control' characteristics

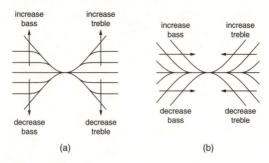

Figure 63 (a) bass; **(b)** treble

'Tilt' control, in which a single knob varies the relative levels of bass and treble (see Figures 65 and 66).

Compressors
Definitions:

1. A **compressor** is a device which reduces the dynamic range of an audio signal by a controllable amount with no significant waveform distortion.

2. **Threshold.** The output level of a compressor is the same as the input level up to the threshold.

3. **Compression ratio.** This is defined as the ratio:

$$\frac{\text{increase in input level}}{\text{increase in output level}} \text{ above the threshold}$$

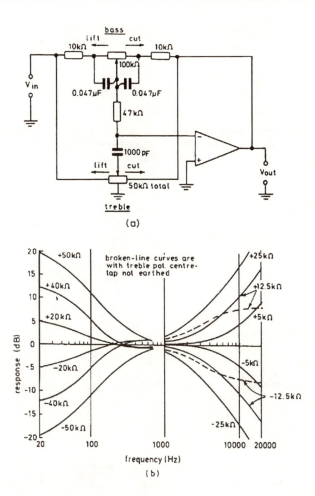

Figure 64 Circuit and response of Baxendall tone controls

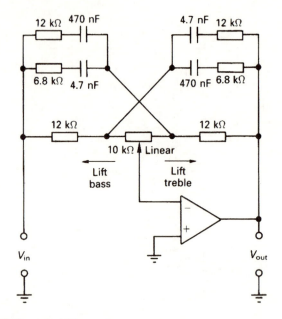

Figure 65 Circuit for 'tilt' control

4. **Limiting.** A very high (ideally infinite) compression ratio. In practice compression ratios higher than about 20:1 are regarded as limiting. On some units the threshold is raised by a fixed amount, typically 8 dB, when 'limiting' is selected.

Figure 67 illustrates the action of what is termed a 'hard-knee' device. Operators sometimes prefer a less abrupt onset of compression. 'Soft-knee' compression is shown in Figure 68.

Methods of compressor control

The control signal ('side-chain') may be used in a 'backwards' or a 'feed-forwards' mode – or possibly both, as illustrated in Figures 69 and 70.

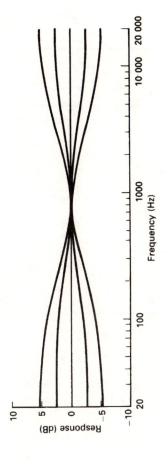

Figure 66 'Tilt' control frequency response

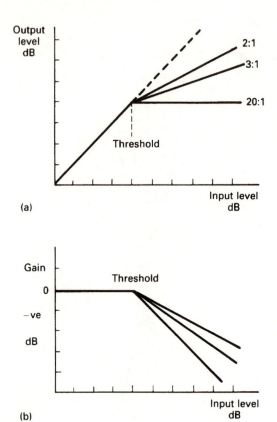

Figure 67 Input/output characteristics of a compressor

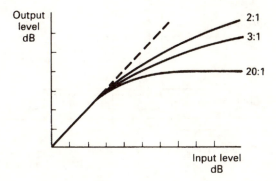

Figure 68 'Soft-knee' compression

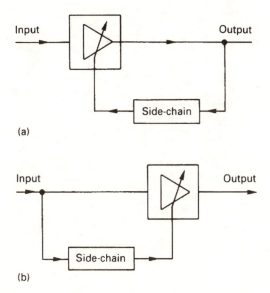

Figure 69 (a) Feed-backwards operation; **(b)** Feed-forwards operation

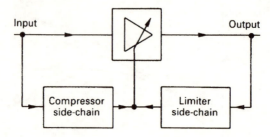

Figure 70 Compressor/limiter with two side-chains

Expanders

These increase the dynamic range. Figure 71 shows the basic block diagram and Figure 72 shows a typical input/output characteristic.

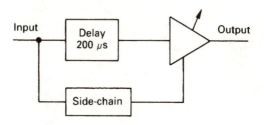

Figure 71 Basic diagram of an expander

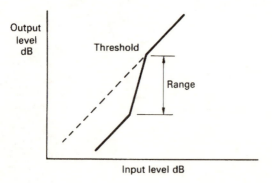

Figure 72 Input/output characteristic of an expander

The expansion ratio is defined as

$$\frac{\text{change in input level in dB}}{\text{change in output level in dB}} \text{ below the threshold}$$

Noise gates

This term is applied to expanders when the expansion ratio is high – 20:1 or more.

16 Analogue recording and reproduction

Formulae

Magnetomotive force, m.m.f is given by

$$\text{m.m.f.} = NI$$

where N is the number of coil turns and I is the current.

The units of m.m.f are ampere turns/meter (At/m)

The intensity or field strength is the total force acting per unit length I of the magnetic circuit. It is called the magnetizing force H.

$$H = \text{m.m.f.}/l = NI/l \, (\text{AT/m})$$

The magnetizing force causes a flux Φ in the magnetic circuit.

The flux density B is the flux per unit area. The unit is the tesla $(= \text{webers/m}^2)$.

$$\Phi = \text{m.m.f.}/S$$

where S is the reluctance of the circuit.

Figure 73 shows a typical BH loop (hysteresis loop).

Magnetic tape – typical characteristics

Base material – Mylar

Base thickness – 25 to 40 μm

Coating, commonly ferric oxide (Fe_2O_3) in the form of needle-shaped particles roughly 0.6 to 1 μm in length and less than one-tenth of that in diameter.

Coating thickness – 10 to 15 μm.

Tape transfer characteristic

Figure 74 shows the distortion that would arise if corrective measures (bias) were not applied.

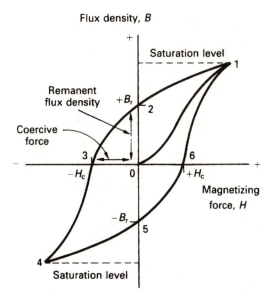

Figure 73 Hysteresis loop

Tape head data

Typical head gaps for 38 cm/s (15 in/s) tape speeds:

     Erase head: ~ 100 μm

     Record head: ~ 20 μm

     Replay head: ~ 5–10 μm or less

Recorded wavelengths

The recorded wavelength is the length on the tape of one cycle of the recorded signal.

$$\lambda_r = \frac{v}{f} \text{ metres}$$

where λ_r = recorded wavelength
          v = tape speed
          f = recorded frequency

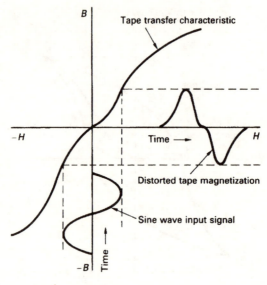

Figure 74 Tape transfer characteristic

Standard tape widths

Tape width		Typical use
(mm)	**(in.)**	
3.8	5/32	Cassettes
6.25	1/4	Mono full-track, 1 or 2 track, stereo
12.5	1/2	Stereo mastering, 4 track
25	1	4-track mastering, 8 track
50	2	16 track, 24 track

In the table below the recorded wavelengths are given in cm.

f	Tape speed (cm/s) / (in/s)			
	4.75	**9.5**	**19**	**38**
(Hz)	**1⅞**	**3¾**	**7½**	**15**
30	0.16	0.32	0.63	1.27
60	0.08	0.16	0.32	0.63
125	0.04	0.08	0.15	0.32
250	0.02	0.04	0.08	0.15
500	9.5×10^{-3}	0.02	0.04	0.08
1 k	4.8×10^{-3}	9.5×10^{-3}	0.02	0.04
2 k	2.4×10^{-3}	4.8×10^{-3}	9.5×10^{-3}	0.02
4 k	1.2×10^{-3}	2.4×10^{-3}	4.8×10^{-3}	0.01
8 k	0.6×10^{-3}	1.2×10^{-3}	2.4×10^{-3}	5×10^{-3}
16 k	0.3×10^{-3}	0.6×10^{-3}	1.2×10^{-3}	2.3×10^{-3}

Comparison of hysterisis loops for tape material and head cores

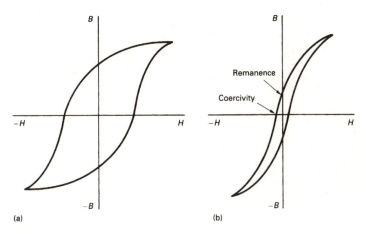

Figure 75 Hysteresis loops for **(a)** 'hard' magnetic material – tape oxide; **(b)** 'soft' – head core material

Bias

Bias frequency
This should be about 4–6 times the highest recorded signal – i.e. around 100 to 150 kHz. Lower bias frequencies may result in undesirable intermodulation products. For example, if the bias frequency were 30 kHz and the highest audio frequency were 18 kHz then there will be sum and difference intermodulation frequencies of 48 kHz and 12 kHz, the latter falling into the audible range.

Optimum bias setting
This is represented graphically in Figure 76.

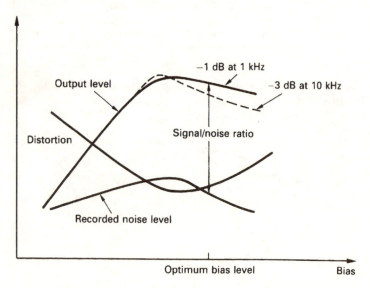

Figure 76 Illustrating optimum bias setting

Replay amplifier equalization

Used to compensate for losses caused by the gap length, the thickness of the tape coating and in the pole pieces, etc.

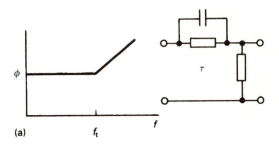

(a)

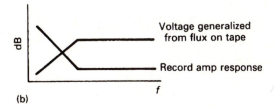

(b)

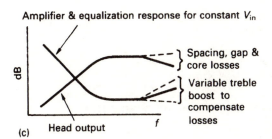

(c)

Figure 77 (a) Equivalent circuit of a standard replay response; **(b)** replay head and amplifier response for an ideal head; **(c)** real replay head and amplifier responses

Standard equalizations in IEC and NAB standards

These are given as the relevant time constants τ, where

$$\tau = \frac{1}{2\pi RC}$$

Tape speed	IEC1	IEC2 and NAB
9.5 cm/s; 3/75 in/s	-	50 Hz; 1800 Hz 3150 μs; 90 μs (NAB only)
19 cm/s; 7.5 in/s	0 Hz; 2240 Hz (∞, 70 μs)	50 Hz; 3150 Hz (3150 μs; 50 μs)
38 cm/s; 15 in/s	0 Hz; 4500 Hz (∞, 35 μs)	50 Hz; 3150 Hz (3150 μs; 50 μs)
76 cm/s; 30 in/s	(IEC1 not used) 0 Hz; 4500 Hz (∞, 35 μs)	AES 1971 0 Hz; 9000 Hz (∞ μS; 17.5 μs)

Head track formats

See Figure 78.

Cassettes

Track layout. See Figure 79.

Mechanical adjustments of a tape head

See Figure 80.

Level

The tone section of a tape is usually recorded at 700 Hz or 1 kHz. The exact frequency is not usually critical. The exact level is, however, important. It is taken as the 0 dB reference but if it is not the reference fluxivity required the example on page 121 shows how to make a correction.

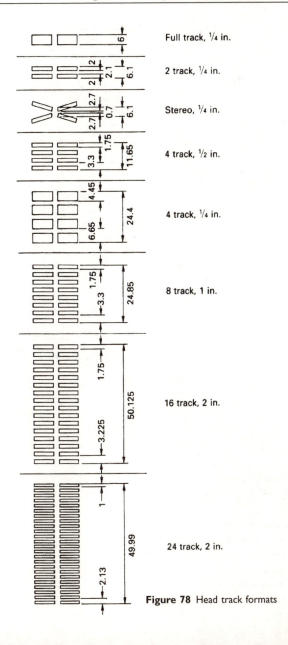

Figure 78 Head track formats

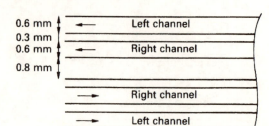

Figure 79 Audio cassette tape

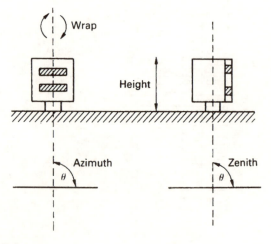

Figure 80 Illustration of azimuth and zenith adjustments

Assume that the machine needs to be aligned to the NAB characteristic at 320 nWb/m for 0 VU = +4 dBm. The reference fluxivity on the tape is 200 nWb/m. The reference fluxivity difference is calculated from

$$\text{Difference (dB)} = 20 \log_{10} \frac{\text{desired reference fluxivity}}{\text{reference fluxivity on tape}}$$

In the example given above:

$$\text{Difference (dB)} = 20 \log_{10} \left(\frac{320}{200}\right) \approx 4 \text{ dB}$$

17 Analogue noise reduction

Noise masking
(See also Section 3 and Figure 8)

Noise masking is an essential part of good noise reduction systems.
Figure 81 illustrates sine-wave masking at 65 dBA SPL.

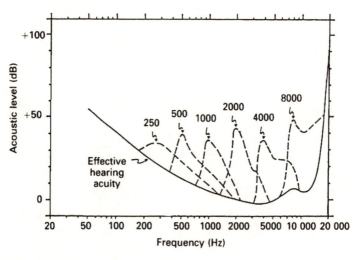

Figure 81 Sine-wave masking at 65 dBA SPL (courtesy of Ehmer (1959))

Elementary Compander

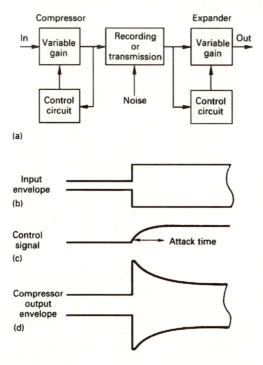

Figure 82 (a) Elementary compander, **(b–d)** the dynamic response of its compressor

Constant-slope devices

Characteristics are shown in Figure 83.

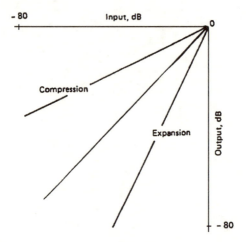

Figure 83 Characteristics of constant-slope noise reduction companders

These devices are independent of input level. As a result:

1. The dynamic characteristics also have to be independent of level.

2. Compression continues down to no signal.

3. All frequencies in the input signal are processed equally.

Bilinear devices

These have a constant gain at low levels, a constant but different gain at high levels, with an intermediate range in which gain changes occur.

It is important with bilinear devices that the decoder receives the correct level signal.

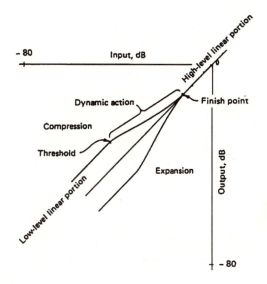

Figure 84 Bilinear compression and expansion

Practical noise reduction systems

Dolby A system

This gives about 10 dB of noise reduction.

High-level signals which are well above the noise are not processed. The audio spectrum is split into four bands:

Band 1: low-pass at 80 Hz;

Band 2: band-pass 80 Hz to 3 kHz;

Band 3: high-pass at 3 kHz;

Band 4: high-pass at 9 kHz.

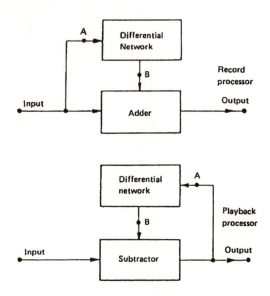

Figure 85 Block diagram of Dolby A noise reduction

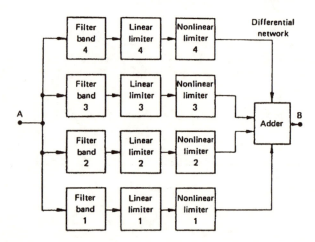

Figure 86 Dolby A differential network. This is the same for coding and decoding

Dolby B system

Widely used to reduce the effects of tape hiss on cassettes this gives about 10 dB of noise reduction in the higher frequencies where tape hiss is most troublesome.

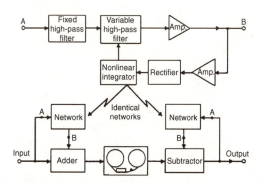

Figure 87 Block diagram of Dolby B noise reduction

The system operates on one frequency band but this has a variable range and can be thought of as a variable pre-emphasis of fixed magnitude. See Figure 88.

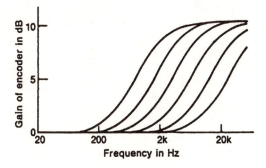

Figure 88 Sliding-band responses of Dolby B noise reduction

Dolby C system

This uses two sliding bands, similar to Dolby B but extending two octaves lower. About 20 dB of noise reduction is achieved.

Dolby SR ('spectral recording') system

Used with a good analogue tape recorder a dynamic range of more than 90 dB can be obtained.

Dolby S-type system

A development of the SR system, this employs two high-frequency stages giving 12 dB of noise reduction above 400 Hz and a single low frequency band providing 10 dB of NR below 200 Hz. It can give up to 24 dB of noise reduction on cassette tape, with a dynamic range of about 85 dB.

dbx systems

These are wide-band constant-slope companders. Types I and II have a 2:1 compression ratio, type 321 has a 3:1 ratio.

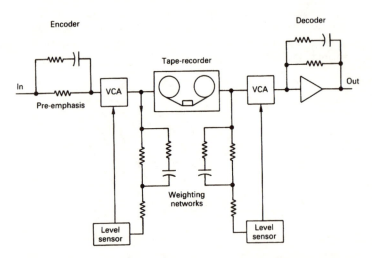

Figure 89 Block diagram of dbx noise reduction systems dbx I. The VCA is preceded by a fixed pre-emphasis in the form of a 12 dB shelf rising from 400 to 1600 Hz.

dbx II. This system is intended for home recording and is similar to type I but includes a band-pass filter to reduce the response to signals outside the range 30 Hz to 10 kHz.

dbx 321. This is a professional system designed for use only with analogue satellite links. It differs from dbx I mainly in having a constant compression ratio of 3:1.

telcom c4. This uses a constant slope of 1.5:1 and four frequency bands as follows:

Band 1: 35 to 215 Hz;

Band 2: 215 to 1450 Hz;

Band 3: 1450 to 4800 Hz;

Band 4: 4800 to 16 000 Hz.

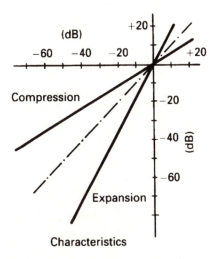

Figure 90 Characteristics of the telcom c4 noise reduction system

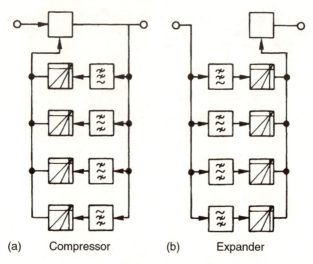

(a)   Compressor   (b)   Expander

Figure 91 telcom c4 system:- **(a)** compressor and **(b)** expander

VHS Hi-Fi

A conventional 2:1 constant-slope compander is used with pre-emphasis in the signal path and frequency weighting in the control path.

18   Compact disc

CD parameters

Disc diameter	120 mm
Rotation speed	568–228 r.p.m. (at 1.4 m/s)
	486–196 r.p.m. (at 1.2 m/s)
Playing time (max)	74 min
No. of tracks	20 625
Track spacing	1.6 μm
Lead-in diameter	46 mm
Lead-out diameter	116 mm
Total track length	5300 m
Linear velocity	1.2 or 1.4 m/s

The CD track
See Figure 92.

Stages in the 'cutting' of a CD
See Figure 93.

1. Glass plate is polished for maximum smoothness.

2. Photo-resist coating applied.

3. Coating is exposed to modulated laser beam.

4. Coating is developed.

5. Surface is silvered to protect 'pits'.

6. Surface is nickel plated to make metal master.

7. Metal master is used to make 'mother' plates.

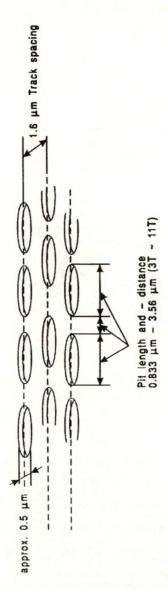

Figure 92 Track and pit dimensions

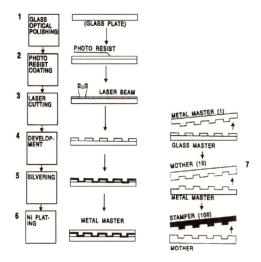

Figure 93 Stages in the production of CD masters and mothers

Error compensation: detection, correction and concealment

Detection: finding out whether there is an error;

Correction: it may be possible to correct completely for the error;

Concealment: if correction is not possible then the effects of the error may be made inaudible, for example by repeating the previous correct sample.

These functions are illustrated in Figure 94.

CRC error detection (Cyclic Redundancy Check)

A bit stream of n bits can be represented as a polynomial with n terms. 11010101 may be written as:

$$M(x) = 1x^7 + 1x^6 + 0x^5 + 1x^4 + 0x^3 + 1x^2 + 0x^1 + 1x^0$$

$$= x^7 + x^6 + x^4 + x^2 + 1$$

Another polynomial $G(x)$ is chosen and in the decoder $M(x)$ is divided by $G(x)$ to give a quotient $Q(x)$ and a remainder $R(x)$.

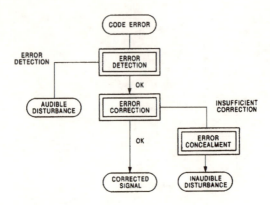

Figure 94 Treatment of errors

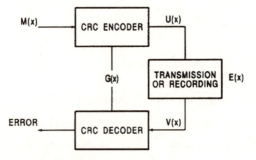

Figure 95 The CRC checking principle

A new message $U(x)$ is then generated so that $U(x)$ can always be divided by $G(x)$ to produce a quotient with no remainder. If there is a remainder then an error has occurred.

With a 16-bit system the detection probability is 99.9985 per cent.

Error concealment. Three types are illustrated in Figure 96.

The faulty samples may be:

   (a) muted;

   (b) the previous sample(s) may be held and repeated;

or (c) an arithmetical operation can be carried out to calculate
        the correct value of the faulty sample.

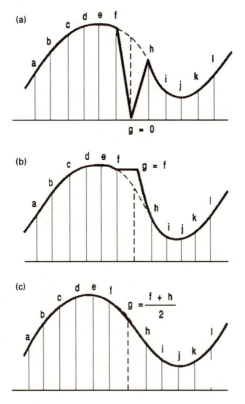

Figure 96 Methods of error concealment

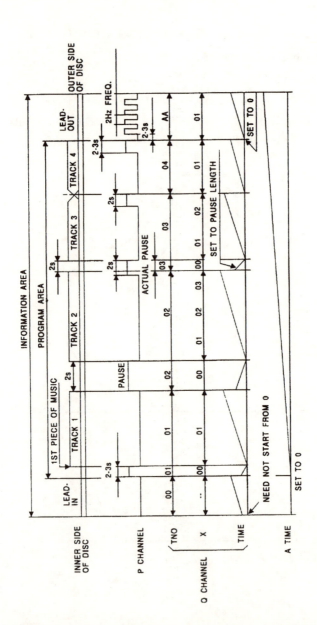

Figure 97 The timing of the P and Q subcodes

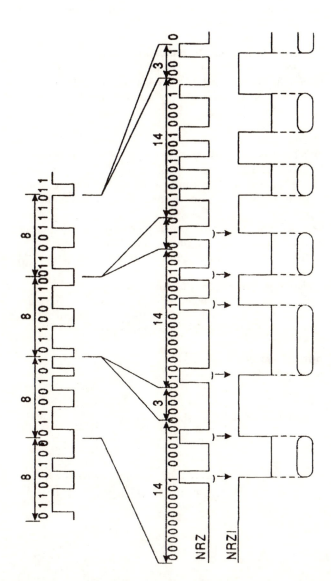

Figure 98 Timing of EFM encoding and merging bits

'Burst' errors

For example, caused by holes in the reflecting surface of the CD (or by 'drop-outs' in digital tape recording).

CIRC – Cross Interleave Reed–Solomon code

The principle is essentially that samples are scrambled to a particular code. The effects of a burst error after descrambling are thus dispersed and can be dealt with by standard error detection and correction methods.

The P and Q subcodes

The P subcode is a music track separator flag. It is normally 0 during music and the lead-in track but is 1 at the start of each selection. It can be used for simple search systems. In the lead-out track it switches between 0 and 1 in a 2-Hz rhythm to indicate the end of the disc.

The Q subcode is used for more sophisticated control purposes; it contains data such as track number and time.

The timing of the P and Q subcodes is shown in Figure 97.

EFM (eight-to-fourteen modulation) coding

This converts each 8-bit symbol into a 14-bit symbol and reduces the required bandwidth, reduces the signal's d.c. content and adds extra synchronizing information (see Figure 98).

CD optical system

See Figures 99, 100.

Tracking

Three possible conditions are shown in Figure 101.

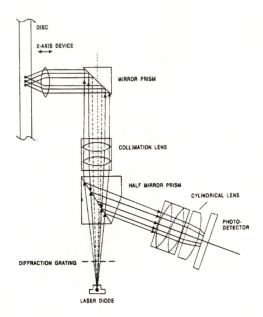

Figure 99 Three-beam optical pickup

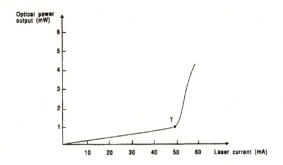

Figure 100 Characteristics of a typical injection laser diode

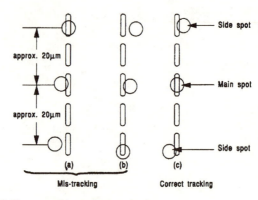

Figure 101 Three possible tracking conditions

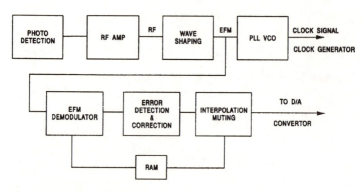

Figure 102 Signal decoding in a CD player

19 Digital audio tape

Introduction

Since the first commercially available digital audio recording systems appeared in 1979, there has been a proliferation of digital tape formats. Many of the early formats are now obsolete but their recorded tapes still exist and need to be recognized. Most formats have been manufacturers' proprietary developments, becoming recognized as standards as more manufacturers agree to support the format. Reel-to-reel digital recording was dominated by the DASH (Digital Audio Stationary Head) and PD (ProDigi) formats, the latter now unsupported. Both accommodated a family of sub-formats, many of which never materialized as commercial products.

The most widely accepted format is R-DAT, or DAT as it is more commonly referred to. While most manufacturers of DAT equipment have kept closely to the standard, commercial and user pressures have created variants to extend applications and technical specification. Some changes such as timecode have been incorporated within the DAT specification, most deviations have not, and so compatibility is not always certain.

In some recent formats, where proprietary rights have been more closely maintained, manufacturers supporting the same format have added non-compatible functionality while maintaining full compatibility at a digital audio track level which makes defining the features of the digital format, manufacturer- and machine-dependent.

It is also necessary to be aware of third party processors being used in conjunction with standard digital multitrack recorders to increase the basic digital audio specification. This normally requires the use of two standard 16-bit digital audio tracks to store a 20 or 24-bit word length within the tape format, reducing the audio track capacity and removing practical format compatibility. This process has been most common within the DTRS digital format.

Format notes

DAT: Consumer DAT recorders only record at 48 kHz sampling rate while professional machines will record at 44.1 and 48 kHz. Both machines replay 44.1 and 48 kHz. The other optional sampling frequencies are now less common.

Deviations from DAT format by manufacturer, achieved by increase in transport speed:

- Pioneer 96 kHz sampling rate option

- Tascam 24-bit word option

- Sonosax 2 or 4 tracks with 44.1/48/88.2/96 kHz sampling rate

DASH HR. DASH multitrack machines are compatible on a track-to-track basis, i.e. a 24-track tape replays on a 48-track while a 24-track will replay 24-tracks of 48-track tape. DASH HR (High Rate) machines are not compatible in the same way and, as different techniques are employed, nor are they compatible between manufacturers' machines.

PD. The ProDigi format is not being supported but there are a lot of machines still in use. There are three variants in the 2-track standard that support modified specifications, i.e. 20-bit or 96 kHz. There is also a 16-track ½-inch version of the multitrack format.

ADAT Types I and II. Type I carries timecode recorded in the subcode area, can record and replay 20-bit digital audio, can record at 44.1 kHz, and carries an analogue track for cue/editing. Type II machines can handle all formats but backwards compatibility is restricted by the bit rate used.

DTRS. DTRS HR version has been announced handling 24-bit digital signals. HR format machines will also replay standard DTRS format tapes.

Nagra D. Machine will record as programmed within the Nagra D format. Can handle up to 24-bit word on tape but choice of A/D converters limits analogue input. Programming option for two or four audio track operation.

Summary of specifications of tape formats in current use

Format	Tape type	Audio channels	Sampling frequency	Word length	Notes
DAT	DAT cassette	2	32/44.056/44.1/48 kHz	16-bit	High spec deviations in format
DASH	½" open reel	24	44.1/48 kHz	16-bit	
DASH	½" open reel	48	44.056/44.1/48 kHz	16-bit	
DASH HR	½" open reel	24/48	44.1/48 kHz	24-bit	Different DASH HR standards
PD	¼" open reel	2	44.1/48(96) kHz	16/20-bit	No model handles all digital specs
PD	1" open reel	32	44.1/48 kHz	16-bit	
ADAT I	S-VHS	8	48 kHz	16-bit	44.1 kHz sampling rate fails in varispeed range
ADAT II	S-VHS	8	44.1/48 kHz	16/20-bit	Includes analogue track for editing
DTRS	Hi8 cassette	8	44.1/48 kHz	16-bit	DTRS HR 24-bit just announced
Nagra D	Open reel	2/4	32/44.1/48 kHz	18/24-bit	Helical scan heads
Sony 1630	U-matic cassette	2	44.056/44.1 kHz	16-bit	CD mastering processor – still in use

Sony PCM1630. The last processor developed for use with U-matic video recorder in CD mastering applications. Still in use although declining.

Head azimuth system
See Figure 103.

DAT cassette discriminating holes
See Figure 107 and the table on page 147.

Tape path
See Figures 108–110.

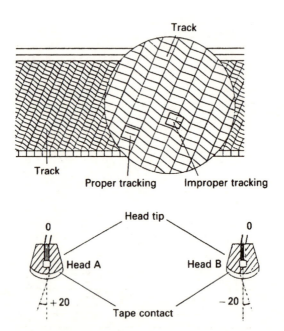

Figure 103 DAT head azimuth system

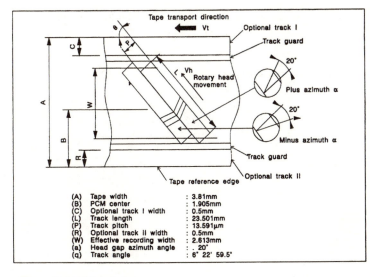

	Tape width	: 3.81mm
(A)	Tape width	: 3.81mm
(B)	PCM center	: 1.905mm
(C)	Optional track I width	: 0.5mm
(L)	Track length	: 23.501mm
(P)	Track pitch	: 13.591µm
(R)	Optional track II width	: 0.5mm
(W)	Effective recording width	: 2.613mm
(a)	Head gap azimuth angle	: . 20°
(q)	Track angle	: 6° 22' 59.5'

Figure 104 DAT track format

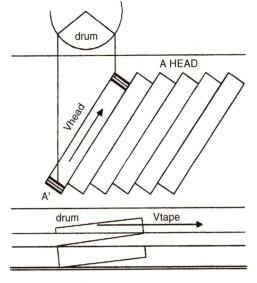

Figure 105 DAT drum position relative to the track

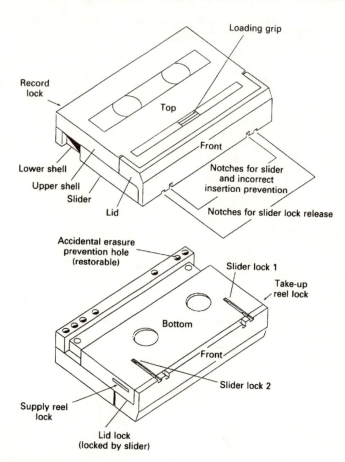

Loading grip

Record lock

Top

Front

Lower shell

Upper shell

Slider

Lid

Notches for slider and incorrect insertion prevention

Notches for slider lock release

Accidental erasure prevention hole (restorable)

Slider lock 1

Take-up reel lock

Bottom

Front

Slider lock 2

Supply reel lock

Lid lock (locked by slider)

Figure 106 The DAT cassette

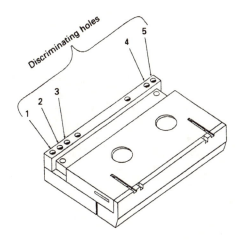

Figure 107 Discriminating holes

Settings of the discriminating holes in DAT cassettes

Hole 1	Hole 2	Hole 3	
0	0	0	Metal tape or equivalent/tape thickness 13 μm
0	1	0	Metal tape or equivalent thin tape
0	0	1	Track pitch 1.5 × /tape thickness 13 μm
0	1	1	Track pitch 1.5 × /thin tape

Hole 4	
1	Prerecorded music tape
0	Non-prerecorded music tape (standard tape for recording)

Hole 5	
1	Recording not possible
0	Recording possible

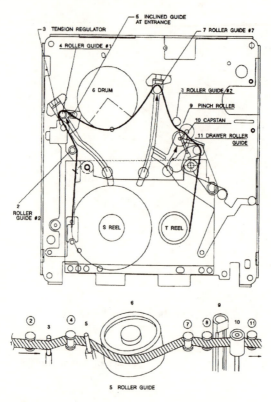

Figure 108 DAT tape path: threaded (———); unthreaded (– – – –)

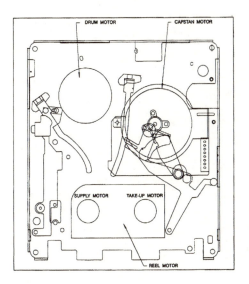

Figure 109 The positions of the DAT motors

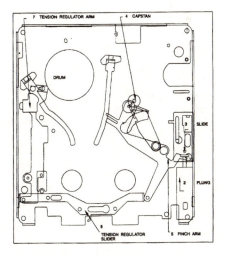

Figure 110 The DAT regulator arm assembly

Main DAT specifications

(a) Record/playback mode

Item	Standard	Option 1	Option 2	Option 3
Channel no. (CH)	2	2	2	4
Sampling frequency (kHz)	48	32	32	32
Quantization bit number (bit)	16	16	12	12
		(linear)	(non-linear)	
Transmission rate (MB/s)	2.46	2.46	1.23	2.46
Modulation system		8–10 conversion		
Correction system		Double Reed–Solomon code		
Cassette size (mm)		73 × 54 × 10.5		
Recording time (min)	120	120	240	120
Tape type		Metal powder		
Tape thickness (μm)		13 ± 1		
Tape speed (mm/s)	8.15	8.15	4.075	8.15
Track pitch (μm)		13.591		
Track angle		6 ° 22′ 59.5″		
Drum rotation speed (r.p.m.)		2000	1000	2000
Relative speed (m/s)		3.133	1.567	3.133
Head azimuth angle (°)		± 20		

(b) Pre-recorded tape (playback only)

Item	Normal track	Wide track
Channel no. (CH)	2	4
Sampling frequency (kHz)		44.1
Quantization bit number (bit)	16	16
Transmission rate (MB/s)		2.46
Modulation system		8–10 conversion

(b) Pre-recorded tape (playback only) *continued*

Item	Normal track	Wide track
Correction system	Double Reed–Solomon Code	
Cassette size (mm)	73 × 54 × 10.5	
Recording time		
(min)	120	80
Tape type	Oxide tape	
Tape thickness (μm)	13 ± 1	
Tape speed (mm/s)	8.15	12.225
Track pitch (μm)	13.591	20.41
Track angle	6 ° 23′ 29.4″	
Drum rotation speed		
(r.p.m.)	2000	
Relative speed (m/s)	3.133	3.129
Head azimuth angle (°)	± 20	

MiniDisc©

This system is becoming increasingly popular in the professional field, where its compactness and ease of use lends it well to locations sound recordings of speech.

There are two types of disc:

1. pre-recorded which are in effect small CDs and are played in a similar way to them;

2. recordable, which are magnetic. In recording the laser heats a very small area of the disc to a temperature above the Curie point (about 185°C), allowing a magnetic head to record the digital data on the disc.

   In the replay mode a much lower laser power in used. The reflected laser beam's polarization is rotated by the magnetic field of the recorded signal and this rotation is detected by a suitable optical process.

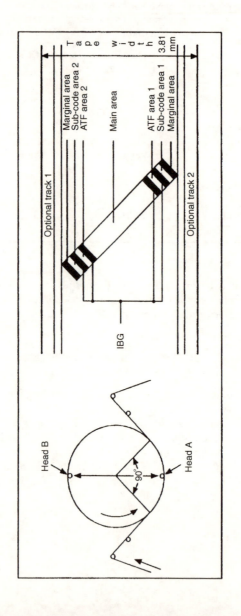

Figure 11.1 Track format. ATF: automatic track finding; IBG, inter-band gap

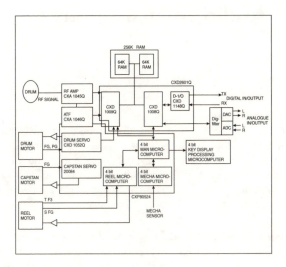

Figure 112 Block diagram of a DAT system

Data:

Disc diameter	64 mm
Disc thickness	1.2 mm
(Cartridge size	72 × 68 × 5 mm)
Track pitch	1.6μm
Scanning velocity	1.2–1.4 m/s
Playing and recording time	74 minutes max (148 minutes in mono)
Sampling frequency	44.1 kHz
Coding	ATRAC (Adaptive Transform Coding)
Laser wavelength	780 nm
Laser power:	
Recording	2.5–5 mW
Replay	0.5 mW approx

20   Audio measurements

Weighting curves

The French CCIR 468–4 is universally adopted for measurements of noise in audio systems. It is shown in Figure 113. The 'unweighted' response curve is needed to avoid the effects of inaudible components which may be present. The A-weighting curve using similar scales is shown in Figure 114 for comparison. The tables give precise values and tolerances.

The use of special dB suffices in audio engineering

Note: A suffix applied to the expression dB is contained in brackets only where it denotes a reference quantity, e.g. dB (mW) expresses a power measurement compared to a reference quantity of 1 mW.

dBu – informal abbreviation of dB (0.775 V)

dBm – informal abbreviation for dB (mW)

dBA – weighted sound pressure level in accordance with BS EN 60651: 1994 (SLOW dynamic)

dBq – audio system noise level (unweighted) using quasi-peak equipment conforming to CCIR 468–4

dBqp – audio system noise level (weighted) using quasi-peak equipment conforming to CCIR 468–4

Audio system noise measurements

Quasi-peak noise measurement equipment conforming to CCIR-4 (see list of standards) is now generally employed for measurements of noise in audio systems. Weighted measurements are made through the curve shown in Figure 113 while the band-pass response for unweighted measurements is also incorporated to eliminate the effect of inaudible components. For expression of readings, use of the quasi-peak dynamic response means that neither dBu nor dBm is appropriate. The standard accordingly specifies the use of dBq

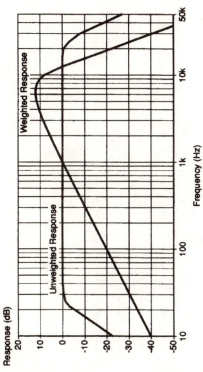

Hz	dB	Tolerance
31.5	-29.9	±2
63	-23.9	±1.4
100	-19.8	±1.0
200	-13.8	±0.85
400	-7.8	±0.7
800	-1.9	±0.35
1k	0	±0.2
2k	+5.6	±0.5
3.15k	+9.0	±0.5
4k	+10.5	±0.5
5k	+11.7	±0.5
6.3k	+12.2	0
7.1k	+12.0	±0.2
8k	+11.4	±0.4
9k	+10.1	±0.6
10k	+8.1	±0.8
12.5k	0	±1.2
14k	-5.3	±1.4
16k	-11.7	±1.6
20k	-22.2	±2.0
31.5k	-42.7	+2.8 −∞

Figure 113 CCIR 468–4 noise weighting curves

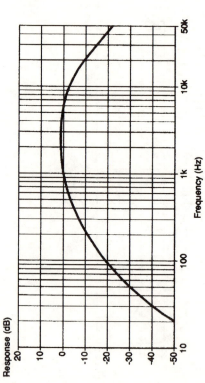

Hz	dB	Tolerance
20	-50.5	±2.0
40	-34.6	±1.0
80	-22.5	±1.0
100	-19.1	±0.5
200	-10.9	±0.5
400	-4.8	±0.5
800	-0.8	±0.5
1k	0	0
1.6k	+1.0	±0.5
2.5k	+1.3	±0.5
4k	+1.0	±0.5
8k	-1.1	±0.5
10k	-2.5	±0.5
12.5k	-4.3	±1.0
20k	-9.3	±1.0
50k	-22.0	±1.0

Figure 114 A weighting curve (IEC179)

and dBqp for unweighted and weighted measurements respectively. On professional equipment noise measurements are made at points where the audio signal peaks to the equivalent of +8 dBu, while the normal line-up tone level of 0 dBu at 1 kHz gives a reading of 0 dBq/0 dBqp on the noise meter.

Weighting curves

Programme level measurements
At the output of professional mixing equipment there is normally a standard programme level meter; this may be either of two patterns:

(a) VU (volume unit) meter. Although this is basically just a moving-coil meter fitted with a bridge rectifier, the ballistics of the meter are very closely specified and only instruments conforming to the IEC/ASA standard should be used (see list of standards). Correct use of the VU meter requires special training and experience.

(b) PPM (peak programme meter). The basic instrument consists of a moving-coil meter having very closely specified ballistics combined with electronic processing to give rapid pointer rise-time and slow fall-back. There are two instruments, standarized by the IEC (see list of standards) and having significantly different characteristics. The 'type 1' (DIN) version has a longer scale and an Integration Time of 5 ms, whereas the 'type 2a' (UK) version has a scale length of 24 dB and an Integration Time of 10 ms. The 'type 2a' specification additionally calls for a 'preferred display meter', covering electrical, dynamic and scale marking features; it is generally used on broadcasting equipment throughout the UK. (The type 2b PPM only differs in respect of the EBU scale.)

Standards relating to audio measurement

Programme level measurement
IEC 60268–10 (2nd edition): 1991/BS 6840: Part 10: 1991
Sound System Equipment Part 10: Methods for specifying and measuring the characteristics of peak programme level meters (Definitive information on PPMs – Type 1 (DIN) and Type 2a/b (UK/EBU).)

IEC 60268–17: 1990/BS 6840: Part 17: 1991
Sound System Equipment, Part 17: Methods for specifying and measuring the characteristics of standard volume indicators (Replaces ANSI/ASA C16.5:1954 – original VU meter specification.)

IEC 60268–18 (1st edition): 1995/BS 6840: Part 18: 1996
Sound System Equipment, Part 18: Peak programme level meters: Guide for digital audio level meter

ITU/R Report BS 292 (was CCIR 292–2) *Use of IEC type 1 PPM*

ITU/R Report BS 820–1: 1994 (was CCIR 820–1) *Comparison of readings using VU meter and PPM*

Audio system noise measurement
ITU/R Recommendation BS 468–4: 1994, *Measurement of Audio-frequency Noise Voltage Level in Sound Broadcasting* (formerly CCIR REC 468–4: 1990) Note: with the exception of the original edition, (CCIR REC 468.1970 which retained RMS noise measurement) previous editions had slightly different tolerances but were essentially the same.

AES presentation, CCIR/ARM, A Practical Noise Measurement Method, Audio Engineering Society, 60th Convention, Los Angeles (1978).

21 Digital equipment

Artificial reverberation
The three main elements are:

1. all-pass filter;

2. comb filter;

3. finite impulse response filter.

See Figures 115–118.

Digital mixing and filtering
See Figure 119.

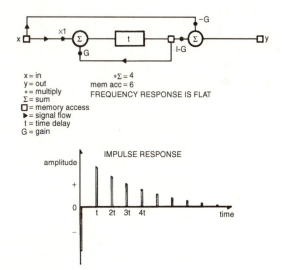

Figure 115 All-pass filter

COMB FILTER

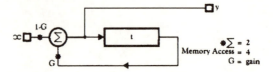

FREQUENCY RESPONSE

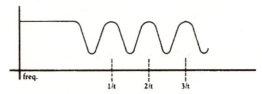

IMPULSE RESPONSE

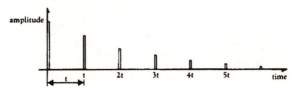

Figure 116 Comb filter

Frequency Response depends on Tap Spacing and Amplitude

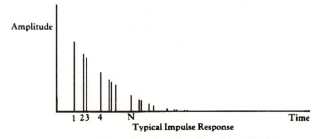

Decay Pattern depends on the values of the * Coefficients.

Tapped delay line F.I.R.

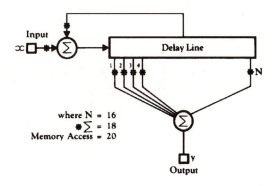

where N = 16
* $\sum$ = 18
Memory Access = 20

Figure 117 Finite impulse response (FIR) filter

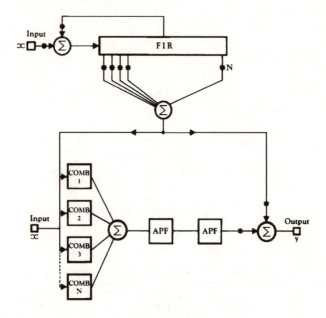

Figure 118 Reverberator with FIR, comb and all-pass filters

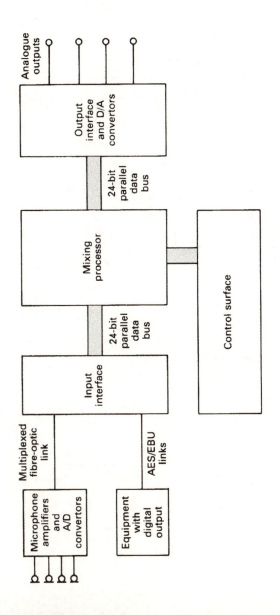

Figure 119 Simplified diagram of a digital mixing console

Basic components of a digital signal processor

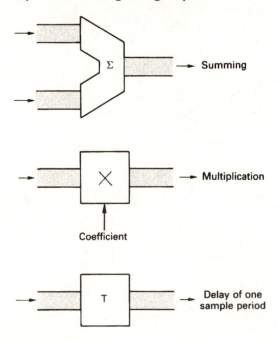

Figure 120 Basic components

Figure 121 shows a simple two-channel digital mixer. The fader settings are converted by the analogue/digital converters into coefficients. The digital channel samples are multiplied by these coefficients, this process being equivalent to a change in the level of the channel sample. The resulting bit stream may be up to 32 bits wide and care must be taken to avoid distortion at low levels when this is converted back to 16 bits.

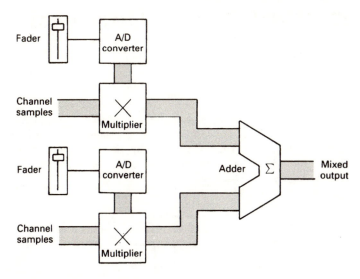

Figure 121 Two-channel digital mixer

Digital filters

These have the advantage over analogue filters that steep slopes can be achieved accurately and with time stability (analogue filters with slopes of more than about 18 dB/octave require critical component values), and phase linear filters are relatively easy to design.

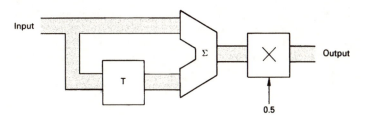

Figure 122 Simple digital filter

22 MIDI (Musical Instrument Digital Interface)

This is a serial interface with start and stop bits at the beginning and end of each byte transmitted. The MIDI standard specifies a baud rate of 31.25 k. (The baud rate is the maximum number of bits per second.) A tolerance of ±1 per cent is specified for the clock frequency.

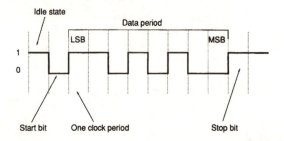

Figure 123 The format of a MIDI message byte. LSB = least significant bit; MSB = most significant bit

MIDI hardware interface
In Figure 124 UART stands for 'universal asynchronous receiver/ transmitter'. The IN connector receives data from other devices, the OUT connector carries data which the device has generated and the THRU connector is a relay of the data present at the IN connector. The THRU connector allows a number of MIDI devices to be 'daisy-chained' together.

Figure 125 illustrates the daisy-chaining of MIDI devices.

MIDI message data format
The status byte contains information about the channel number to which the message applies. It denotes which receiver the message

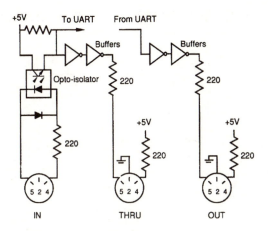

Figure 124 The MIDI hardware interface

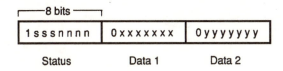

Figure 125 A MIDI daisy-chain

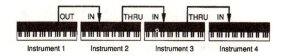

Figure 126 MIDI message date format

MIDI data messages

Message	Status	Data 1	Data 2
Channel-specific messages			
NOTE OFF	&8n	Note number	Velocity
NOTE ON	&9n	Note number	Velocity
POLYPHONIC AFTERTOUCH	&An	Note number	Pressure
CONTROL CHANGE	&Bn	Controller no.	Data
14-bit controllers MSByte	&Bn	01 (modulation wheel)	Data
(examples)		02 (breath controller)	Data
		04 (foot controller)	Data
		05 (portamento time)	Data
		06 (data entry slider)	Data
		07 (main control)	Data
14-bit controllers LSByte	&Bn	21 (mod wheel etc)	Data
7-bit controllers/switches	&Bn	40 (sustain pedal)	}00–3F(off)
(examples)		41 (portamento)	}40–7F(on)
		42 (sostenuto pedal)	}40–7F(on)
		43 (soft pedal)	}40–7F(on)

		60 (data increment)	7F
		61 (data increment)	7F
		62 (non-regulated parameter control)	LSByte
		63 (non-regulated parameter control)	MSByte
		64 (regulated parameter control)	LSByte
		65 (regulated parameter control)	MSByte
		79 (reset all controllers)	7F
Channel modes	&Bn	7A (local)	00 off/7Fon
		7B (all notes off)	00
		7C (omni off)	00
		7D (omni on)	00
		7E (mono)	00
		(No. of channels)	00–OA
		7F (poly)	00
PROGRAM CHANGE	&Cn	Program No.	—
CHANNEL AFTERTOUCH	&Dn	Pressure	—
PITCH WHEEL	&En	LSByte	MSByte

MIDI data messages (continued)

Message	Status	Data 1	Data 2
System messages			
System exclusive:			
SYSTEM EXCLUSIVE START	&F0	Manufacturer ID	Data+++ ...
END OF SYSTEM EXCLUSIVE	&7F	–	
System Common			
SONG POINTER	&F2	LSByte	MSByte
SONG SELECT	&F3	Song number	–
TUNE REQUEST	&F6	–	
MIDI time-code			
QUARTER-FRAME	&F1	Data	–
System real-time			
TIMING CLOCK	&F8		
START	&FA		
CONTINUE	&FB		
STOP	&FC		
ACTIVE SENSING	&FE		
RESET	&FF		

is intended for and also which type of message is to follow –
e.g. a NOTE OFF message. The status byte is given in hexadecimal
form, e.g. &9n, n being the channel number in hexadecimal form.

The tables on pages 168–170 give MIDI data messages.

Sequencers

A sequencer stores MIDI information taken from one or more MIDI
inputs with the ability to send it out at a later time from one or
more MIDI outputs. During the storage time editing and other data
manipulation may take place. See Figure 127.

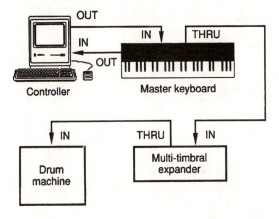

Figure 127 An example of a sequenced MIDI system

Figure 128 shows a more complex system in which timecode from a tape machine is used for synchronization.

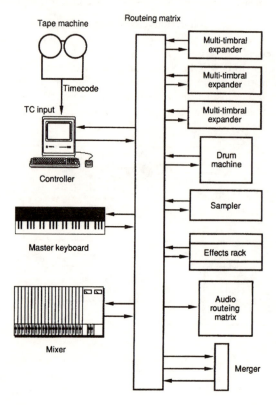

Figure 128 A complex MIDI system using time-code from a tape machine for synchronization

23   Studio air-conditioning

Air quality
Potential pollutants:

1. reduced oxygen and increased carbon dioxide from people;

2. water vapour, cigarette smoke;

3. formaldehyde from studio furnishings.

Ventilation requirements
To maintain carbon dioxide at the maximum recommended level of 0.5 per cent by volume the minimum ventilation requirement is 0.8 l/s for seated persons.

Dilution of body odours to acceptable levels requires a fresh air ventilation rate of 8 l/s for sedentary non-smoking occupants. When heavy smokers are present the rate should be at least 32 l/s.

Thermal comfort

Temperature
A person seated and at rest produces approximately 115 W of heat (75 per cent by radiation and convection, 25 per cent by evaporation).

The comfort of an individual depends on:

1. the ambient air temperature;

2. the mean radiant temperature;

3. the average air velocity;

4. air humidity;

5. clothing worn;

6. activity undertaken.

The thermal index most widely used in the UK is denoted by the resultant temperature, t_{res}, and is given by

$$t_{res} = \frac{t_r + (t_{ai} \sqrt{10v})}{1 + \sqrt{10v}}$$

where t_{ai} is the inside air temperature (°C), t_r is the mean radiant temperature (°C) and v is the mean air velocity (m/s).

The recommended resultant temperature range for sedentary occupations, as in studios, is 19–23°C, if the relative humidity is between 40 per cent and 70 per cent and the air velocity is less than 0.1 m/s.

When the air velocity is less than 0.1 m/s the above expression simplifies to

$$t_{res} = \tfrac{1}{2}t_r + \tfrac{1}{2}t_{ai}$$

Humidity

Relative humidity, RH is defined as

$$RH = \frac{\text{actual vapour pressure}}{\text{saturation vapour pressure}} \, 100\%$$

See table opposite.

For most studios the RH should be between 40 per cent and 70 per cent. If the RH is less than about 40 per cent static electrical charges can develop.

RH from wet-and-dry bulb thermometer readings

Depression of wet bulb (°C)	Dry bulb temperature (°C)							
	14	16	18	20	22	24	26	28
0.5	95	95	95	96	96	96	96	96
1.0	90	90	91	91	92	92	92	93
1.5	85	85	86	87	87	88	88	89
2.0	79	81	82	83	83	84	85	85
3.0	70	71	73	74	76	77	78	78
3.5	65	67	69	70	72	73	74	78
4.0	60	63	65	66	68	69	71	72
4.5	56	58	61	63	64	66	67	69
5.0	51	54	57	59	61	62	64	65
5.5	47	50	53	55	57	59	61	62
6.0	42	46	49	51	54	56	58	59
6.5	38	42	45	48	50	53	54	56
7.0	34	38	41	44	47	49	51	53
7.5	30	34	38	41	44	46	49	51
8.0	26	30	34	37	40	43	46	48
8.5	22	26	30	34	37	40	43	45
9.0	18	23	27	31	34	37	40	42
9.5	14	19	23	28	31	34	37	40
10.0	10	15	20	24	28	31	34	37

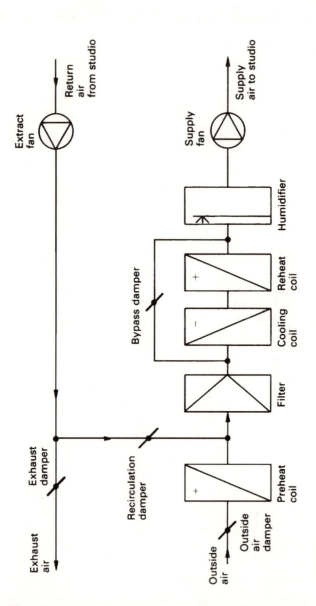

Figure 129 Diagram of a typical air conditioning system

24 Distribution of audio signals

Electromagnetic telephones

Basic characteristics and functions:

1. Hybrid. The telephone is a four-wire device but needs to be converted to two-wires for connection to the exchange.

2. Side-tone. Some of the transmitted signal is fed into the receiver.

3. Power feed. The transmitter has to be fed with d.c. but the receiver needs to be isolated from this.

4. Impedance matching. The a.c. impedance of the telephone needs to be matched with the impedance of the line. The telephone must be able to cope with a wide variation in line impedances.

5. Protection – from extraneous voltages, power line surges, lightning.

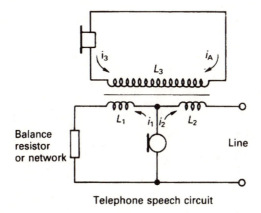

Telephone speech circuit

Figure 130 Telephone speech circuit

The speech circuit. Figure 130 shows the use of a hybrid transformer for use with electromagnetic telephones. The current from the receiver is split into two components, I_1 and I_2. These produce currents I_3 and I_4 respectively. These nearly cancel out – the residual current provides the side-tone.

Basic signalling
This is shown in outline in Figure 131.

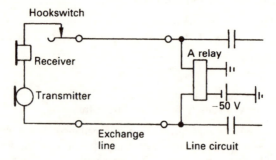

Figure 131 Basic signalling for electromagnetic telephones

Electronic telephones
The essential requirements are similar to those of electromagnetic telephones:

1. provide two-wire to four-wire conversion;

2. extract power to operate the transducers and integrated circuits;

3. set send and receive loudness gains;

4. provide side-tone;

5. match impedances.

Two-wire to four wire conversion is illustrated in the bridge shown in Figure 132.

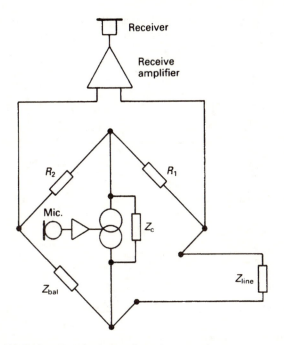

Figure 132 Bridge circuit for two- to four-wire conversion

For perfect balance of the bridge in Figure 132

$$\frac{Z_{\text{line}}}{Z_{\text{bal}}} = \frac{R_1}{R_2}$$

Transducers in electronic telephones

1. Microphone. Generally of the electret type.

   The electret material carries a permanent charge equivalent to a d.c. potential of 100 V.

2. Receivers. Either moving coil or rocking armature. See Figure 134.

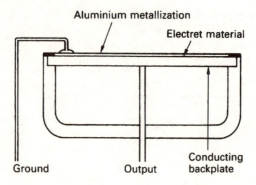

Figure 133 Electret microphone

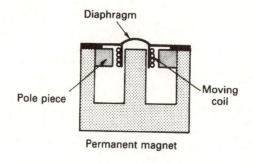

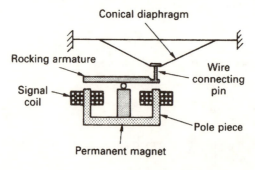

Figure 134 Moving coil receiver (top), rocking armature (below)

Dual-tone multifrequency signalling (DTMF)

Sixteen audio frequencies are arranged in a four-by-four matrix. Digits are represented by two out of eight frequency combinations. For example the digit 6 is formed by 770 Hz plus 1447 Hz.

Low frequencies (Hz)	High frequencies (Hz)			
	1209	1336	1447	1633
697	1	2	3	A
770	4	5	6	B
852	7	8	9	C
941	*	0	#	D

The tones are usually 50 ms in duration with intervals of 40 ms between tones.

ISDN (integrated services digital network)

The aim is to provide voice and non-voice services in the same network. Two basic delivery methods have been defined for ISDN.

1. A basic rate service running at 144 Kbits/s, giving 2 × 64 Kbits/s channels and 16 Kbits/s signalling channel.

2. A primary rate service running at 2.048 Mbits/s yielding 30 × 64 Kbits/s channels for signalling and synchronization.

25 Radio propagation

The broadcast range of frequencies

Medium wave (MF): 526.5–1606.5 kHz
Very high frequency (VHF Band 2):
nominally 88–108 MHz
Short wave (SW) or high frequency (HF):
3900–4000 kHz (75 m band)
5950–6200 kHz (49 m band)
7100–7300 kHz (41 m band)
9500–9775 kHz (31 m band)
11 700–11 975 kHz (25 m band)
15 100–15 450 kHz (19 m band)
17 700–17 900 kHz (16 m band)
21 450–21 750 kHz (13 m band)
25 600–26 100 kHz (11 m band)

Effective isotropically radiated power (EIRP)

This is the product of power fed to the actual antenna multiplied
by the antenna gain with respect to an isotropic antenna which
radiates uniformly in all directions. Figure 135 shows how the
power density varies with distance for an EIRP of 1 kW.

Reference field strength

This is 1 μV/m. Other fields are usually quoted as dB with respect
to this reference. Figure 136 shows the relationship between dB
μV/m and the field in μV/m.

The ionosphere

This stretches from approximately 50 km above the earth. It results
from UV and X-radiation and also high energy charged particles
from the sun. For convenience the ionosphere is divided into the
D,E and F layers, the last splitting at times into F_1 and F_2 layers.
Figure 137 shows the heights of the layers in a simplified way and
also the nature of the variations in height throughout the day.

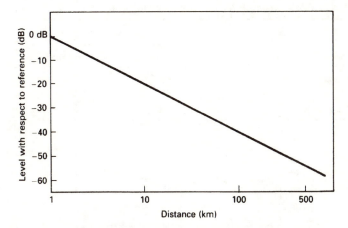

Figure 135 Variation of power density with distance for an EIRP of 1 kW. The reference (0 dB) at 1 km is approximately 7.96×10^{-5} W/m

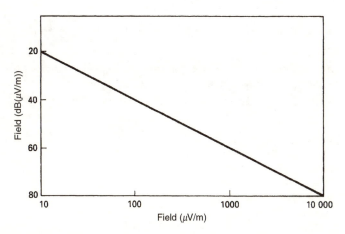

Figure 136 Relationship between field strength and decibel field

Median field strength for f.m. broadcasting in band-2 VHF

Area	Field strength (dB)	
	Monophonic	**Stereophonic**
Rural	48	54
Urban	60	66
Large cities	70	74

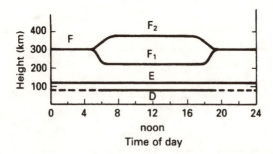

Figure 137 Ionospheric layer heights with diurnal variations (simplified)

Critical frequency (F_{crit}) is the maximum frequency reflected back to earth at vertical incidence from a given layer.

Typical values are:

$$F_{crit}E = 4\ \text{MHz}$$

$$F_{crit}F_1 = 6\ \text{MHz}$$

$$F_{crit}F_2 = 15\ \text{MHz}$$

Properties of the layers

D layer. Approximately 50 to 90 km above the earth. During daytime MF waves are almost completely absorbed by this layer. At sunset absorption decreases rapidly so that skywave reflection from the E layer occurs.

E layer. About 110 km above the earth. MF signals are reflected from it at night time. Lower frequency HF waves can also be reflected in the daytime. Maximum ionization happens in summer at around noon.

Sporadic E (E_s) occurs when clouds of intense ionization form. Abnormal propagation over large distances may then take place.

F layer. Upwards of about 150 km. It splits in the daytime into F_1 and F_2.

F_1 is around 200 km above the earth and has greater ionization than the E layer resulting in higher critical frequencies. $F_{crit}F_1$ peaks in summer.

F_2 is about 300 km above the earth. Ionization is greater than in any of the other layers. $F_{crit}F_2$ is greater in winter than in summer.

26   Digital interfacing and synchronization

The AES/EBU interface
(AES = Audio Engineering Society; EBU = European Broadcasting Union).

This format allows two channels of digital audio to be transferred serially over one balanced interface using RS422 drivers and receivers. Figure 138 shows the hardware. This allows the two channels to be transferred over distances up to 100 m.

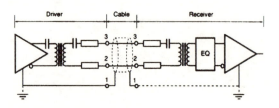

Figure 138 AES/EBU hardware interface

The format of the 'sub-frame' is shown in Figure 139.

The 'Sync' bits can be one of three patterns which identify which of the two channels the sample represents or marks the start of a new channel status block. 'Aux' carries additional data. See also Figure 140.

The Sony–Philips digital interface (SPDIF)
This is very similar to the AES/EBU interface with only subtle differences from it.

Sony digital interface (SDIF)
The common version is the SDIF-2 which is used mainly for the transfer of audio data from SONY professional digital audio

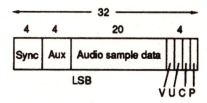

Figure 139 AES/EBU subframe format

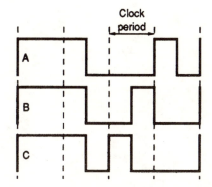

Figure 140 AES/EBU sync patterns. A is the start of the A-channel sub-frame, B is the start of the B-channel sub-frame and C is the start of a new channel status block

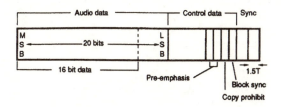

Figure 141 SDIF-2 data format

equipment, particularly the PCM-1610 and 1630, and also on some semi-professional equipment (see Figure 141).

Mitsubishi interfaces

Mitsubishi ProDigi format machines use an interface similar to SDIF but not compatible with it. Separate interconnections are used for each audio channel. 'Dub A' and 'Dub B' are 16-channel interfaces found on multi-track machines; Dub A handling tracks 1–16 and Dub B handling tracks 17–32. The pin assignments are shown below.

Dub A:

Pin	Function
1,18	Ch.1(+/−)
2,19	Ch.2(+/−)
3,20	Ch.3(+/−)
4,21	Ch.4(+/−)
5,22	Ch.5(+/−)
6,23	Ch.6(+/−)
7,24	Ch.7(+/−)
8,25	Ch.8(+/−)
9,26	Ch.9(+/−)
10,27	Ch.10(+/−)
11,28	Ch.11(+/−)
12,29	Ch.12(+/−)
13,30	Ch.13(+/−)
14,31	Ch.14(+/−)
15,32	Ch.15(+/−)
16,33	Ch.16(+/−)
34,35	Bit clock(+/−)
36,37	WCLK(+/−)
38,39	Rec A(+/−)
40,41	Rec B(+/−)
17,50	GND

Dub B:

Pin	Function
1,18	Ch.17(+/−)
2,19	Ch.18(+/−)
3,20	Ch.19(+/−)

4,21	Ch.20(+/−)
5,22	Ch.21(+/−)
6,23	Ch.22(+/−)
7,24	Ch.23(+/−)
8,25	Ch.24(+/−)
9,26	Ch.25(+/−)
10,27	Ch.26(+/−)
11,28	Ch.27(+/−)
12,29	Ch.28(+/−)
13,30	Ch.29(+/−)
14,31	Ch.30(+/−)
15,32	Ch.31(+/−)
16,33	Ch.32(+/−)
17,50	GND

Dub C

Pin	Function
1,14	Left(+/−)
2,15	Right(+/−)
5,18	Bit clock(+/−)
6,19	WCLK(+/−)
7,20	Master clock(+/−)
12,25	GND

Yamaha cascade interface

This is often used with Yamaha digital audio equipment to allow a number of devices to be operated in cascade. The interface terminates in an 8-pin DIN-type connector and carries two channels of 24-bit audio data over an RS422-standard differential line.

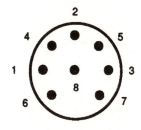

Figure 142 Yamaha interface connector pins

Pin connections are:

Pin	Function
1	WCLK +
2	GND
3	Audio data −
4	WCLK −
5	Audio data +
6	20 μH coil to GND
7	20 μH coil to GND
8	GND (in), ENABLE (out)

The 20 μH inductors on Pins 6 and 7 are for suppression of r.f. interference.

Timing for synchronous signals

More than one reference signal may be used for locking the sampling-rate clock of a digital audio device. In a large system it is vital that all devices remain locked to a common sampling rate reference clock and there are AES recommendations for this. Input signal frame edges must lie within ± 25% of the reference signal's frame edge, and output signals within ± 5%, although tighter accuracy than this is preferable. (See Figure 143.)

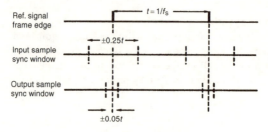

Figure 143 Timing stipulations for synchronous signals

27   Ultrasonics

Piezoelectric transducers

Natural materials: quartz, SiO_2.

Synthetic materials include certain plastics and ceramics containing lead zirconate and titanate (PZT). The latter have to be electrically polarized by placing them in a strong electric field. For this reason they are known as 'ferroelectrics' by analogy with the magnetization of a ferromagnet. Four constants characterize the performance and efficiency of piezoelectric transducers:

1. Bandwidth, Q

$$Q = \frac{f_{fun}}{(f_a - f_b)}$$

where f_{fun} = the fundamental resonant frequency, and f_a and f_b are the frequencies at which the transducer output amplitude falls to $1/\sqrt{2}$ of the amplitude at f_{fun}.

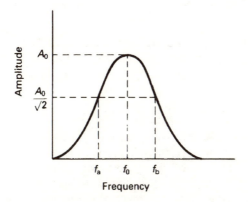

Figure 144 Graph of amplitude versus frequency, yielding the Q factor

2. Electromechanical coupling coefficient, k

$k = \sqrt{}$(fraction of mechanical energy that goes to produce an electric field in the piezo)

or

$k = \sqrt{}$(fraction of electrical energy that goes to produce mechanical energy)

3. $g = \dfrac{\text{electric field in piezo slice}}{\text{stress applied to piezo faces}}$

(The electric field is in V/m, the stress is in Pa.)

4. $d = \dfrac{\text{charge on piezo face}}{\text{force applied to wafer face}}$

(Charge is in C, force is in N.)

or

$d = \dfrac{\text{change in length of wafer face}}{\text{voltage applied to faces}}$

(Units are m and V, respectively.)

In general:

Good receiver:	low d,	high g
Good transmitter:	high d,	low g

Electromagnetic acoustic transducers (EMAT)

Detection using an EMAT is illustrated in Figure 145 and Figure 146.

Magnetostrictive transducers

A change in the dimensions of the material results from the application of a magnetic field. The amount of change depends on the material and its internal structure. Some materials such as Turfinol (Fe, Tb and Py) can expand by up to 1 per cent.

Ferromagnetic materials can be ultrasonically tested for defects by making use of their own magnetostrictive properties, as shown in Figure 147.

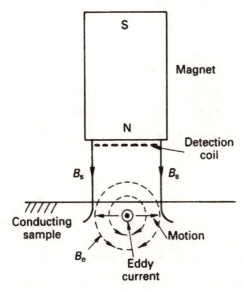

Figure 145 In-plane motion detection. B_s is the static magnetic field due to the magnet, B_e is the varying magnetic field caused by the eddy current

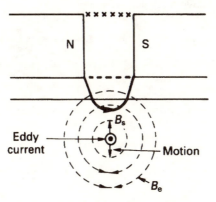

Figure 146 Out-of-plane motion detection with an EMAT

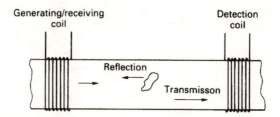

Figure 147 Reflection and transmission of ultrasonic waves in a magnetostrictive sample

Reasons for the use of ultrasonics in industry

1. Non-destructive – the sample is not affected and there may not even need to be contact between the sample and the ultrasound generator and detector.

2. Environmentally safe and clean.

3. Easy to use. Operators of ultrasound equipment need little training and many processes can be automated.

4. Accurate. Surface displacements of as little as 0.02 nm can be detected using laser interferometry. Thickness measurements down to 10 μm are possible.

28   Radio studio facilities

General purpose studios
This category covers most on-air and preparation studies, news, interviews, presentation and announcement studios.

Layout. Figure 148 shows a typical small studio complex.

The sound lobbies are essential for providing acoustic separation between studio and control room.

Acoustics
A reverberation time of 0.4 to 0.5 is generally acceptable for this type of studio.

The mixing desk
Number of channels: often about eight for a small speech-only studio. 16–24 channels is more typical.

Important desk facilities include:

> Prefade listen (PFL) on each channel or group – especially important when there are incoming live contributions.
>
> Equalization.
>
> Auxiliary outputs – at least two, for artificial reverberation, foldback or clean feeds to contributors, etc.
>
> Outside sources (OS). A mixing desk in a general purpose studio should be capable of handling a variety of outside sources, including telephone calls via the public network.

Telephone balance units (TBU)
Such units must be able to:

> 1. hold a call when it is diverted from a normal telephone handset to the mixing desk;

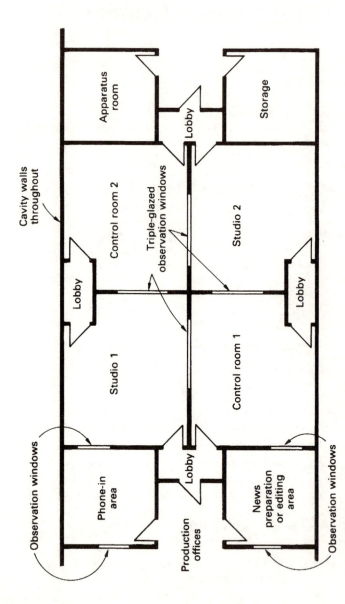

Figure 148 Typical small studio complex

2. terminate the incoming line with the correct impedance;

3. isolate electrically the telephone network from the studio equipment;

4. separate the send and receive signals from the two-wire telephone circuit for connection to the desk;

5. provide automatic gain control of the send and receive signals;

6. provide a voice-over circuit to reduce the level of the telephone call when the studio presenter speaks, thus reducing unwanted colouration.

In connection with connections to the public telephone system some broadcasters insert an audio delay of several seconds ('Profanity Delay') to allow a presenter time to react to unacceptable incoming language.

Communications

The following are typical requirements in a general purpose studio:

1. a quick and fool-proof talk-back system;

2. a dedicated talk-back to tape;

3. an intercomm. linking with, for example, a newsroom;

4. cue lights – often green. A pre-arranged set of signals will be used.

Transmission safeguards

Important ones are:

1. the ability to disable any line-up oscillators which may be used for setting signal levels;

2. the routeing of talkback to studio headphones only when microphones are live and not to studio loudspeakers or to the desk output.

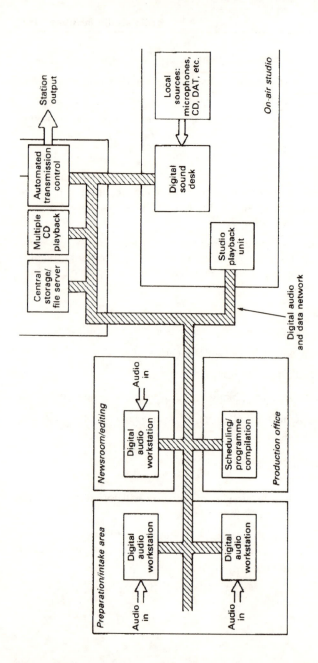

Figure 149 Semi-automated radio station complex

Recording and playback formats:

1. Vinyl disc players – still likely to be needed.

2. Open-reel ('quarter-inch') analogue tape machines.

3. CD players.

4. Cartridge players (e.g. for jingles).

5. Cassette recorders.

6. DAT machines.

7. MiniDisc players.

8. Digital mass storage, such as networked hard sick systems. The latter allow semi-automated systems (see Figure 149).

Rock music studios

These are similar in almost every way to commercial recording studios.

Acoustics

Often very short reverberation times although more flexible working arrangements are often provided. These may include 'live' and 'dead' areas.

Typical control room facilities.

1. Separate apparatus room for multi-track tape machines, desk power supplies, digital processing racks etc.

2. Distribution of computer and MIDI data signals around the area.

3. The mixing desk is likely to be complex, probably providing computer-assisted mixing and assignable facilities. There will be enough auxiliary sends to allow for the use of a wide range of out-board equipment.

Listening levels in rock studios

SPLs may well exceed 100 dBA. Current UK legislation requires precautionary action to be taken when levels are greater than 85 dBA.

Classical music studios

Acoustics
Reverberation times in the region 1.2 to 2.0 s, as flat as possible between 50 Hz and 10 kHz. Background noise levels should be very low – say 25 dBA.

Live broadcasts
Serious music studios may frequently be used for live broadcasts and this needs to be borne in mind when specifying facilities.

Radio drama studios

Acoustics
To allow satisfactory manipulation of acoustics and perspective requires a range of conditions:

1. dead room. Typical reverberation time is 0.1 s;

2. 'live' area with a reverberation time in the range 0.5 to 0.8 s;

3. effects area with, e.g., purpose-built kitchen, telephone booth;

4. narrator's booth.

A typical drama studio layout is shown in Figure 150.

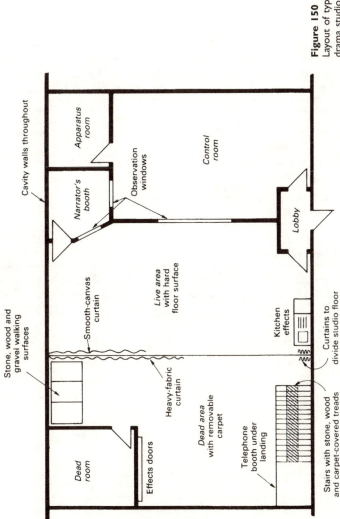

Figure 150
Layout of typical radio drama studio

Cavity walls throughout

Apparatus room

Narrator's booth

Observation windows

Control room

Lobby

Stone, wood and gravel walking surfaces

Smooth-canvas curtain

Live area with hard floor surface

Kitchen effects

Curtains to divide studio floor

Heavy-fabric curtain

Dead room

Effects doors

Dead area with removable carpet

Telephone booth under landing

Stairs with stone, wood and carpet-covered treads

29 Connectors

Balanced circuits

Connector type	Signal (+ or 'hot')	Signal (− or 'cold')	Earth (screen)
XLR style	Pin 2*	Pin 3*	Pin 1
3-terminal jack (e.g. PO style or so-called 'stereo' jack, 6.3 mm or miniature, wired for a single balanced circuit)	Tip	Ring	Sleeve
3-pin DIN (as fitted to some microphones)	Pin 1	Pin 3	Pin 2

Unbalanced circuits

Connector type	Signal (+ or 'hot')	Earth (screen) and signal (− or 'cold')
XLR style	Pin 2*	Pins 3* and 1
Phono	Pin	Sleeve
2-terminal jack (6.3 mm or miniature)	Tip	Sleeve
3-terminal jack (6.3 mm or miniature)	Tip	Ring and sleeve
3-terminal jack (6.3 mm or miniature) wired for stereo	L: Tip R: Ring	Sleeve

* XLR pins 2 and 3 are reversed by some organizations and in some, mainly American, microphones. When connected to equipment using the other standard this can cause a phase-reversal in balanced circuits and a possible loss of signal in unbalanced ones.

DIN connectors

Note that there can be variations depending on the manufacturer of the equipment.

DIN 180° (5 pin)		
Microphones	L: Pin 1 R: Pin 4	Pin 2 — Pins 3 and 5 may carry the polarizing voltage
Tape recorder inputs	L: Pin 1 R: Pin 4	Pin 2 —
Tape recorder outputs:		
(Low impedance)	L: Pin 1 R: Pin 4	Pin 2 —
(High impedance)	L: Pin 3 R: Pin 5	Pin 2 —

European voltage standardization

For many years the UK domestic supply voltage has been based on a 415/240 volt range. From 1 January 1995 the permitted range changed to a 400/230 volt supply. In effect the new range of permitted variations is larger and almost entirely covers the older range.

User	Supply voltage and permitted variations before 1 January 1995	Supply voltage and permitted variations after 1 January 1995
Most domestic (single phase supplies)	240 volts 225.6 − 254.4 v	230 volts 216.2 − 253 v
Most commercial or industrial (three phase supplies)	415/240 volts 390.1 − 439.9 v	400/230 volts 376.0 − 440 v
Other (split phase supplies)	480/240 volts 451.2 − 508.8 v	460/230 volts 432.4 − 506 v

30 Public address data

100 volt line loudspeaker systems
(Data reproduced by courtesy of Klark Teknik.)

Load impedance versus power dissipation		Load impedance versus power dissipation	
ohms	watts	ohms	watts
10	1000	1.0 k	10
20	500	1.11 k	9
25	400	1.25 k	8
33.3	300	1.43	7
40	250	1.67	6
50	200	2.00 k	5
83	120	2.50 k	4
100	100	3.33 k	3
111	90	5.00 k	2
125	80	10.0 k	1
200	50	20.0 k	0.5
250	40	40.0 k	0.25
333	30	80.0 k	0.125
500	20		

Absorption of sound in air for different relative humidities (dB/30 m)

Note that peak absorption for all frequencies occurs between 10 and 20 per cent RH. The peak at 2 kHz occurs when RH is about 10 per cent; at 10 kHz when RH is about 20 per cent.

f (kHz)	RH(%)				
	20	30	50	70	90
2	1.0	0.5	0.2	0.2	0.1
4	2.7	1.6	1.0	0.9	0.8
6	4.8	2.9	1.8	1.5	1.5
8	6.5	4.3	3.0	2.4	2.3
10	8.6	5.9	4.0	3.3	3.0

31　Useful literature

The following list is by no means exhaustive but nevertheless contains books, papers and articles which may be helpful if more information is needed.

The items in the list are put into categories in an attempt to help the reader. However because there is an inevitable overlap in categories the same works sometimes appear more than once.

Hearing

GREEN, D.M., *An Introduction to Hearing*. Earlbaum, Hillsdale, NY (1976)

MOORE, B.C.J., *Introduction to the Psychology of Hearing*. Academic Press, London (1986)

PICKLES, J.O., *An Introduction to the Physiology of Hearing*. Academic Press, London (1986)

ROBINSON, D.W. and DADSON, R.S. A re-determination of the equal loudness relations of pure tones, *British Journal of Applied Acoustics*, **7**, 166–181 (1956)

Transducer

BAXENDALL, P.J., *Loudspeaker and Headphone Handbook*, 2nd edn. Focal Press, Oxford (1994)

BORWICK, J., *Microphones; Technology and Technique*. Focal Press, Oxford (1990)

GAYFORD, M.L., *Electroacoustics, Microphones, Earphones and Loudspeakers*. Newnes-Butterworths, Oxford (1970)

JORDAN, E.J., *Loudspeakers*. Focal Press, London (1963)

NESBITT, A., *The Use of Microphones*, 4th edn. Focal Press, London (1995)

OLSON, H.F., *Acoustical Engineering*. Professional and Audio Journals, Philadelphia, PA (1991)

ROBERTSON, A.E. *Microphones*. Iliffe, London (1963)

Room acoustics

BBC, *Guide to Acoustic Practice*. British Broadcasting Corporation (1990)

BERANEK, L. L. *Music, Acoustics and Architecture*. Wiley, New York (1962)

BISHOP, R.E.D. and JOHNSON, D.C., *Mechanics of Vibration*. Cambridge University Press, Cambridge (1979)

GILFORD, C.L.S., *Acoustics for Radio and Television Studios*. Peter Peregrinus, London (1972)

HARRIS, C.M. (ed), *Handbook of Noise Control*, 2nd edn. McGraw-Hill, New York (1979)

KNUDSEN, V.O., *Architectural Acoustics*. Wiley, New York (1932)

SABINE, P.E., *Acoustics and Architecture*. McGraw-Hill, New York (1932)

TEMPLETON, DUNCAN (Ed.) *Acoustics in the Built Environment*. Butterworth Architecture, Oxford (1993)

Stereo
BLAUERT, J., *Spatial Hearing: the Psychophysics of Human Sound Localisation*. J.S. Allen, MIT Press, Cambridge, MA (1983)

BLUMLEIN, A.D., British Patent Specification 394325, *Journal of the Audio Engineering Society*, **6**, 91–100 (1958)

DAUBNEY, C., Ambisonics – an operational insight, *Studio Sound* (Aug 1982)

MOORE, B.C.J., *Introduction to the Psychology of Hearing*. Academic Press, London (1986)

SNOW, W., Basic principles of stereophonic sound, *Journal of the Society of Motion Picture and Television Engineers*, 61, 567–589 (1953)

Amplifiers and filters
AMOS, S.W., *Radio, TV and Audio Technical Reference Book*. Newnes-Butterworth, London (1977)

BAXANDALL, P.J., Negative-feedback tone control, *Wireless World*, 58, 402–405 (1952)

Limiters and compressors
BEVILLE, M., Compressors and limiters: their uses and abuses, *Studio Sound*, **19**, 28–32 (1977)

DUNCAN, B., VCAs investigated, *Studio Sound and Broadcast Engineering*, **31** (7) 68–62 (1989)

GLEAVE, M.M. and MANSON, W.I., The development of sound-programme limiters in the BBC, *BBC Engineering* **107**, 9–10 (1977)

Analogue recording
ARNOLD, A.P., *Principles of Magnetic Tape Recording (BBC Engineering Training Sheet 45T)* BBC, Evesham (1979)

HAMMOND, P., *Electromagnetism for Engineers*. Pergamon Press, Oxford (1978)

Noise reduction

BERKOVITZ, R., and GUNDRY, K.J., Dolby B-type noise reduction, *Audio Magazine* (Sep/Oct 1973)

DOLBY, R.M., A 20 dB audio noise reduction system for consumer applications, *Journal of the Audio Engineering Society*, **31**, 98–113 (1983)

DOLBY, R.M., The spectral recording process, *Journal of the Audio Engineering Society*, **35**, 99–118 (1987)

EHMER, R.H., Masking of tones v. noise bands, *Journal of the Acoustical Society of America*, **31**, 1253–1256 (1959)

Compact disc

CLIFFORD, M., *The Complete Compact Disc Player*, Prentice-Hall, Englewood Cliffs, NJ (1987)

POHLMANN, K.S., *Principles of Digital Audio*. Macmillan, New York (1985)

SONY SERVICE CENTRE, *Digital Audio and Compact Disc Technology*, 3rd edn. Focal Press, Oxford (1995)

Digital audio tape recording and editing

BORWICK, J. (ed.), *Sound Recording Practice*. Oxford University Press, Oxford (1995)

HOAGLAND, A.S. and MONSON, J.E., *Digital Magnetic Recording*. Wiley, New York

IMMINK, K.A.S., *Coding Techniques for Digital Recorders*. Prentice Hall, Englewood Cliffs, NJ

RUMSEY, F.J., *Digital Audio Operations*. Focal Press, Oxford (1991)

WATKINSON, J., *The Art of Digital Audio*, 2nd edn. Focal Press, Oxford (1994)

MiniDisc

MAES, J., *The MiniDisc*. Focal Press, Oxford (1996)

Digital equipment, etc.

POHLMANN, K.S., *Principles of Digital Audio*. Macmillan, New York (1985)

RUMSEY, F.J. *MIDI Systems and Control*, 2nd edn. Focal Press, Oxford (1994)

SYPHA, *The Nonlinear Video Buyer's Guide*, 5th edn. Sypha, 216A Gipsy Road, London, SE7 9RB (1999)

SYPHA, *The Tapeless Audio Directory*, 7th edn. Sypha, 216A Gipsy Road, London SE7 9RB (1998)

WATKINSON, J., *The Art of Digital Audio*, 2nd edn. Focal Press, Oxford (1994)

Mobile

DEPARTMENT OF TRANSPORT, *Road Vehicles (Construction and Use) Regulations*, current edition. HMSO, London.

Air conditioning

LUFF, M.G., *Air Conditioning for Students*. Technitrade Journals, London (1980)

MILLER, L.M., *Students' Textbook of Heating, Ventilation and Air Conditioning*. Technitrade Journals, London (1976)

Telephony

HILLS, M.T. and EVANS, B.G., *Transmission Systems*. Allen and Unwin, London (1973)

RICHARDS, D.L., *Telecommunications by Speech*. Butterworth, Oxford, (1973)

SMITH, S.F., *Telephony and Telegraphy*. Oxford University Press, Oxford (1969)

Teletext and RDS

MOTHERSOLE, P.L. and WHITE, N.W., *Broadcast Data Systems – Teletext and RDS*. Focal Press, Oxford (1993)

Digital interfacing

RUMSEY, F., *Digital Audio Operations*. Focal Press, Oxford (1991)

RUMSEY, F. and WATKINSON, J., *The Digital Interface Handbook*, 2nd edn. Focal Press, Oxford (1995)

WATKINSON, J., *The Art of Digital Audio*, 2nd edn. Focal Press, Oxford (1994)

Public address

CAPEL, V., *Public Address Systems*. Focal Press, Oxford (1992)

Ultrasonics

DUBOVY, J., *Introduction to Biomedical Electronics*. McGraw-Hill, New York (1978)

FREDERICK, J.R., *Ultrasonic Engineering*. Wiley, New York (1965)

SHUTILOV, V.A., *Fundamental Physics of Ultrasound*. Gordon & Breach, New York (1988)

SZILARD, J., (ed.) *Ultrasonic Testing, Non-Conventional Testing Techniques*. Wiley, Chichester (1982)

Index

A signals, 87, 88
A weighting network, 23
A-B powering, 71
Absorption coefficients, sound, 42–3, 206
Acoustic elements, electrical equivalents, 34
Acoustic resonances, 4
Acoustics, radio studio, 195, 199, 200
ADAT machines, 142
AES/EBU interface, digital audio transfer, 186, 187
After-fade listen (AFL), 100
Air-conditioning, studio, 173–6
Air quality/pollution, 173
Aliasing, 39
All-pass filter, 159
American Standards, 27
Amplitude modulation, 7
Analogue noise reduction, 122–30
Analogue recording and reproduction, 112–21
Analogue sound mixing, 94, 94–103
Artificial reverberation, 159, 162
Audio cassette tape, track layout, 120
Aural monitoring, sound mixing, 101
Automatic track finding (ATF), 151
Azimuth adjustment, tape head, 120

B signals, 87, 88
Baffles, loudspeaker, 81
Balanced/unbalanced circuits, 94
  connectors, 202

Bases, mathematical, 3
Basic principles, 1–11
Bass tip-up *see* Proximity effect
Baxendall tone controls, 105
Beats sensation, 22
Bias frequency, 116
Bias setting, 116
Bilinear companders, 124, 125
Bit rate, 37
British Standards, acoustic noise, 27
Broadcast frequencies, 182

Cardioid microphone, 65–6, 91
Cassette tape, audio, 120
CCIR, 468-4 standard, audio noise, 154, 155
Channel facilities, sound mixing, 97, 100
Circuits, balanced/unbalanced, 94
Classical music studios, 200
Clean feed, 101
Close box baffle, 81
Coincident pair microphones, 88
Comb filter, 160
Compact disc (CD), 131
  cutting stages, 131, 133
  error compensation, 133
  optical system, 139
  signal decoding, 140
  track and pit dimensions, 132
  tracking conditions, 140
Compander:
  bilinear, 124, 125
  constant-slope, 124, 128–9
  elementary, 123

Compression ratio, 104
  limiting, 106
Compressor, 104
  control methods, 106, 109, 110
Connectors:
  balanced circuit, 202
  DIN, 202, 203
  unbalanced circuit, 203
Constant-slope companders, 124,
  128–9
Control room facilities, 199
Crossover networks, 77, 79–81
Cyclic redundancy check (CRC),
  133–4

DASH HR machines, 142
DASH machines, 142
dB suffices, noise levels, 154
dbx compander systems, 128–9
Decibel (unit), 4
Definitions, 1–3
Degrees of freedom, system, 35
Delay, audio, 197
Difference tones, 22
Digital audio tape (DAT), 141
  automatic track finding (ATF),
    151
  block diagram, 153
  cassette, 146
  cassette discriminating holes,
    147
  drum position, 145
  formats, 142, 143, 145, 151
  head azimuth system, 144, 145
  specifications, 150
  tape path, 148, 149
Digital principles, 37–40
Digital signal processor:
  basics, 39, 163, 164, 165
  mixer, 165
DIN connectors, 203
Directivity patterns, microphone,
  59–69
Dither, 38

Dolby A system, 125, 126
Dolby B system, 127
Dolby C system, 128
Dolby S-type system, 128
Dolby SR system, 128
Doppler effect, 14
DTRS machines, 142
Dual-tone multifrequency
  signalling (DTMF), 181
Dynamic range, 94

Ear:
  frequency discrimination, 19
  frequency responses, 19
  structure, 19, 20
Effective isotropically radiated
  power (EIRP), 182
EFM coding, 137, 138
Eight-to-fourteen modulation *see*
  EFM coding
Electret microphone, 180
Electrets, 58
Electrical formulae, 5–6
Electromagnetic acoustic
  transducers (EMAT), 192, 193
Electromechanical relationships,
  31–3
Electrostatic microphones, 58–9
Enclosures, loudspeaker, 82
Equal loudness curve, 19, 21
Equivalent noise level, 24
Error concealment, 135
Error detection, 133–4
  digital signal processing, 39
European voltage standardization,
  204
Expander, 110
Expansion ratio, 111
Exposure times, 25
Eyring and Norris formula, 41

Ferroelectrics, 191
Figure-of-eight microphones, 62–5,
  90

Filters, 159, 160, 161
  digital, 165
Finite impulse response (FIR) filter,
  161
Flux density, 112
Fluxivity adjustment, magnetic
  tape, 118, 121
Frequency, 3, 12, 13
  typical sounds ranges, 16
Frequency difference limen, 19
Frequency modulation, 7

Group modules, sound mixing, 101
Gun microphone, 67, 68, 69

Haas effect, 85
Head track formats, 119
Health and safety regulations, 25
  Action Levels, 25
Hearing damage, 24
Hearing process, 19–22
Helmholtz resonator, 4–5, 82
Horn loading, 82
Humidity requirements, studio, 174
Hypercardioid microphone, 66–7,
  90
Hysteresis loop, 112, 113, 115

IEC standards, acoustic noise, 27
Impedance relationships, 31
Input impedances, 95
Integrated services digital network
  (ISDN), 181
Inter-channel difference, sound
  image position, 86
Interconnections, 96
  see also Connectors
Interference-tube microphone, 67,
  68, 69
International standards, acoustic
  noise, 27
Inverse square law, 14
Ionosphere, 182, 184–5
Italian musical terms, 17–18

Jack connecting plugs, 95, 96
JFMG Ltd, 73
  address, 75

Licensing, transmitters, 75
Lip ribbon microphone, 64
Loudness, 20
  design criteria, 45–6
Loudspeakers, 76–83
  100 volt line systems, 205
  impedance and frequency, 82
  sensitivity, 83
  types, 72–7

M signals, 87, 91
M/S microphones, 88, 91
M/S signals, derivation from A/B,
  88
Magnetic tape, 112
  recorded wavelengths, 113, 115
  standard widths, 114
  transfer characteristic, 112
Magnetomotive force, 112
Magnetostrictive transducers, 192,
  194
Membrane resonance, 4
Microphone transducers, 55–9
Microphones:
  directional response, 54
  electret, 180
  electrical characteristics, 55
  electrostatic, 58–9
  M/S, 88, 91
  physical aspects, 54
  radio *see* Radio microphones
  reliability, 54
  sensitivities, 55, 56
  sound quality, 53
  for stereo, 88–92
  types, 55–9
  variable directivity, 67
  vulnerability, 54
MiniDisc system, 151
Mitsubishi interfaces, 188–9

Mix minus *see* Clean feed
Mixing desk, 195
Modal density, 41
Modulation index, 11
Monitoring systems, sound mixing, 101–103
Mono line channel, 98
Most significant bit (MSB), 38
Moving-coil drive units, loudspeaker, 76
Moving-coil microphones, 55–6
Moving-coil receiver, 180
Multiple driver units, loudspeaker, 77
Musical Instrument Digital Interface (MIDI), 166
  daisy-chaining, 167
  hardware interface, 166
  message data format, 166, 167–71
  sequencer, 171
  synchronization system, 172
Musical scale, 14, 15
Musical terms, Italian, 17–18

Nagra D machines, 142
Noise:
  definitions, 23–4
  design criteria, 45–6
  see also Hearing damage
Noise criteria (NC) curves, 48
Noise gates, 111
Noise level meters, 23, 25–6, 157
Noise levels:
  addition, 26
  effect of different materials, 50–52
  measurement, 23–4, 26, 154, 157
  measurement standards, 24, 158
  practical reduction systems, 125–9
  typical, 28–30
Noise masking, 21, 122
Noise rating (NR) curves, 47
Normalling (interconnections), 96, 97

Ohm's law, 5
Omnidirectional microphones, 59–62
Oversampling, 39

P and Q subcodes, 136, 138
Panpot microphone systems, 92, 93
Parity (error detection), 39–40
Peak programme meter (PPM), 103, 157
Permeability of free space, 3
Phantom power systems, 69
Phase cancellation, 60, 61
Phon (unit), 21
Piezoelectric loudspeaker, 77
Piezoelectric transducers, 191
  bandwidth, 191
  electromechanical coupling coefficient, 192
Playback formats, 199
Polar diagrams *see* Directivity patterns
Power ratios, decibel values, 8
Pre-fade listen (PFL), 100
Pressure units, 12
ProDigi (PD) machines, 142
Profanity delay, 197
Programme level measurements, 157, 158
Proximity effect, 63, 64
Public address systems, 205–206

Quantizing levels, 37

Radio drama studios, 200, 201
Radio microphone frequencies, 72–5
Radio propagation, 182–5
Radio station, semi-automated, 198
Radio studio facilities, 195–201
Rayleigh's formula, 41
Recording formats, 199
Reference field strength, 182

Relationships, useful, 1
Relative humidity, 174, 175
Relative permeability, 3
Replay amplifier equalization, 117, 118
Resistor, colour code, 6
Resonance formulae, 4, 5, 6
Reverberation, artificial, 159, 162
Reverberation time, 41
    effect on sound reduction index, 49
    recommendations, 43, 44
    studio, 199, 200
Ribbon loudspeaker, 77
Ribbon microphone, 57–8
Rock music studios, 199
    listening levels, 199
Rocking armature receiver, 180

S signals, 87, 91
Sabine formula, 41
Safeguards, transmission, 197
Sampling rate, 37
Signal levels, 94
Signal processing, 104–11
Signal-to-noise (S/N) ratios, 38
Solo-in-place, sound mixing, 101
Sone (unit), 20
Sony digital interface (SDIF), 186, 187
Sony PCM1630 machines, 144
Sony–Philips digital interface (SPDIF), 186
Sound absorption, for different relative humidities, 206
Sound image position:
    inter-channel difference, 86
    for stereo microphones, 89
Sound isolations:
    airborne, 49
    building materials, 49–51
Sound level meters *see* Noise level meters
Sound mixing, analogue, 94

Sound pressure level (SPL):
    decibel values, 5, 9, 10
    loudspeaker sensitivity, 83
    reference zero, 19
Sound reduction index (SRI), 49
    effect of reverberation time, 49
    typical values, 51–2
Sound velocity, 12, 13
Sound waves, physics, 12–18
Standard equalizations, IEC and NAB standards, 118
Standing wave patterns, 16
Star quad cable, 94
Stereo effects, 84–92
    M and S signals, 87, 88
    terminology, 87
Stereo line channel, 99
Studio reverberation times, 44
Synchronous signals timing, 190

Tape head adjustments, 120
Tape head data, 113
Telcom c4 noise reduction system, 129, 130
Telephone balance unit (TBU), 195, 197
Telephones:
    electromagnetic, 177–8
    electronic, 178–9
Temperature requirements, studio, 173–4
Threshold, 104
Tilt control, 104, 106, 107
Time of arrival (TOA) difference, 84
Tone control, 104, 105
Transmission line enclosure, 82
Transmission safeguards, 197
Transmitter licensing, 75
TV channel frequencies, 73, 74
Two's complement, 38

Ultrasonics, 191–3
    industrial uses, 194
Unbalanced circuit connectors, 202

Universal asynchronous receiver/
transmitter (UART), 166

Variable directivity microphone, 67
Vented enclosures *see* Helmholtz
resonator
Ventilation requirements, studio, 173
VHF frequencies, 72
VHS Hi-Fi system, 130
Vibration frequencies, 14, 17
Vibration isolation analysis, 36
Visual monitoring, sound mixing,
101

Voltage standardization, European,
202, 204
Volume unit (VU) meter, 157

Wavelength, 3, 12, 13
Weighting curves:
noise measurement, 154, 155, 156
programme level measurement,
157

Yamaha cascade interface, 189–90

Zenith adjustment, tape head, 120

Also available from Focal Press ...

Audio Engineer's Reference Book
Edited by Michael Talbot-Smith

An authoritative volume on all aspects of audio engineering and technology including:
- basic mathematics and formulae
- acoustics and psychoacoustics
- microphones
- loudspeakers
- studio installations, including air conditioning

Compiled by an international team of experts, the second edition has been updated to keep abreast of fast-moving areas such as digital audio and transmission technology. For professionals engaged in the design, manufacture and installation of all types of audio equipment, this reference book will prove an invaluable resource.

'An excellent easy to read compendium of both theory and practice. Clearly the writers are experts, making this a valuable reference publication. It's all in there; from definitions of mass, time and current, to setting up and running broadcast and recording studios.'
Sound & Communications International

1999 • 640pp • 246 x 189mm • Hardback
ISBN 0 240 515285

Focal Press

http://www.focalpress.com

Visit our web site for:

- •The latest information on new and forthcoming Focal Press titles
- •Technical articles from industry experts
- •Special offers
- •Our email news service

Join our Focal Press Bookbuyers' Club
As a member, you will enjoy the following benefits:

- •Special discounts on new and best-selling titles
- •Advance information on forthcoming Focal Press books
- •A quarterly newsletter highlighting special offers
- •A 30-day guarantee on purchased titles

Membership is FREE. To join, supply your name, company, address, phone/fax numbers and email address to:

USA
Christine Degon, Product Manager
Email: christine.degon@bhusa.com
Fax: +1 781 904 2620
Address: Focal Press,
225 Wildwood Ave, Woburn,
MA 01801, USA

Europe and rest of World
Elaine Hill, Promotions Controller
Email: elaine.hill@repp.co.uk
Fax: +44 (0)1865 314572
Address: Focal Press, Linacre
House, Jordan Hill, Oxford,
UK, OX2 8DP

Catalogue
For information on all Focal Press titles, we will be happy to send you a free copy of the Focal Press catalogue:

USA
Email: christine.degon@bhusa.com

Europe and rest of World
Email: carol.burgess@repp.co.uk
Tel: +44 (0)1865 314693

Potential authors
If you have an idea for a book, please get in touch:

USA
Terri Jadick, Associate Editor
Email: terri.jadick@bhusa.com
Tel: +1 781 904 2646
Fax: +1 781 904 2640

Europe and rest of World
Christina Donaldson, Editorial Assistant
Email: christina.donaldson@repp.co.uk
Tel: +44 (0)1865 314027
Fax: +44 (0)1865 314572